A Primer of Invertebrate Learning
THE BEHAVIORAL PERSPECTIVE

Charles I. Abramson

AMERICAN PSYCHOLOGICAL ASSOCIATION • WASHINGTON, DC

Published by
American Psychological Association
750 First Street, NE
Washington, DC 20002

Copies may be ordered from
APA Order Department
P.O. Box 2710
Hyattsville, MD 20784

In the United Kingdom and Europe, copies may be ordered from
American Psychological Association
3 Henrietta Street
Covent Garden
London WC2E 8LU
England

Typeset in New Baskerville and Futura by Easton Publishing Services, Inc., Easton, MD

Printer: Wickersham Printing Company Inc., Lancaster, PA
Cover designer: Berg Design, Albany, NY
Illustrator: Hans & Cassady, Inc., Westerville, OH; figures based on renderings by
    Anna Yuan Miller and Daniela Brunner
Cartoonist: Andy Myer, Maple Glen, PA
Technical/production editor: Valerie Montenegro

Library of Congress Cataloging-in-Publication Data
Abramson, Charles I.
    A primer of invertebrate learning: the behavioral perspective / Charles I. Abramson.
        p.   cm.
    Includes bibliographical references (p.   ) and index.
    ISBN 1-55798-228-7 (acid-free paper)
    1. Invertebrates—Behavior.   2. Invertebrates—Psychology.
3. Learning in animals.   I. Title.
QL364.2.A28 1994
592′.051—dc20

94-4300
CIP

British Library Cataloguing-in-Publication Data
A CIP record is available from the British Library.

Printed in the United States of America
First edition

Contents

Figures

Exhibits and Tables

Preface

I wrote this book for a number of reasons. Not the least of these is my firm belief that the study of invertebrates has a great deal to contribute to the understanding of basic learning processes. Who has not been intrigued by the secret language of bees, or perplexed at the duty-bound lives of worker ants, or amazed (albeit horrified) at the ability of the cockroach to adapt and thrive on anything from potato peels to postage stamps? Invertebrates have inspired fascination for the wonders of the world for thousands of years.

It follows that the study of invertebrate learning is the story of many remarkable successes. Several notable scientists began their careers studying invertebrates: Sigmund Freud examined crayfish for information on nerve cells and fibers; Ivan Pavlov studied shell opening in clams; B. F. Skinner analyzed the geotropic response in the ant. Others have gone on to win Nobel prizes for their work with invertebrates. Still others, like Rachel Carson, have altered the way people care for the environment by demonstrating the hazards of insecticides and pesticides.

Among the many advances made possible by the study of invertebrates are an increase in knowledge of the learning process, a greater appreciation of the plasticity inherent in stereotypic behavior, and the use of training methods to assist bees in the pollination of crops. In addition, simple system researchers have made significant inroads in understanding the cellular mechanisms of behavior.

Although these advances are impressive, a careful look at the field reveals some fundamental problems. One formidable problem is a lack both of adequate teaching aids and of a meaningful knowledge base. There is a real shortage of parametric data, no generally accepted classification of behavior, and little in the way of standardized behavioral techniques.

In addition, the field seems to be influenced by fashions that come and go. Those readers who are familiar with the invertebrate literature of the 1950s and 1960s, in which the behavior of planarians, protozoans, and many other invertebrates was extensively studied, may be puzzled to see this vibrant literature ignored in contemporary reviews. In other cases, early experiments that were used to demonstrate learning often lacked the controls now known to be important, and those studies should be replicated. Much work remains to be done in establishing unifying principles.

My hope is that this book will at least partially address those problems by providing a sound methodological introduction to invertebrate learning that familiarizes readers with basic principles. In order to conduct effective experiments with invertebrates, it is crucial to understand the importance of apparatus design, the need to implement various control procedures, the fine points of behavior analysis, and those nuances of experimental design that can make the difference between success and failure. In the chapters that follow, I have tried to increase the reader's awareness of those concerns.

I have also attempted to give the reader some background information about the data generated, the methods used, and the traditional problems that have been posed in the study of learning and memory. It is important that all of us who conduct learning experiments with invertebrates be familiar with the rich research that has preceded us. In this spirit, I have provided an exhaustive list of references—both for locating published material and for finding individuals who are working in a particular area—in the hope that the interested reader will be encouraged to indulge his or her curiosity and dig deeper.

Finally, I am pleased to offer the reader comprehensive information on invertebrate learning apparatus. My review of learning apparatus is unique in that it includes several Russian techniques that have not yet appeared in the English language. A clever use of apparatus can go far in stretching a meager laboratory research budget. It is my hope that this review will provide practical inspiration for those whose resources are more constricted than their ingenuity.

The study of invertebrate learning is truly interdisciplinary, and I believe that this primer will be useful to undergraduate students in psychology as well as to zoologists, entomologists, biochemists, ethologists, and psychologists. While newcomers to this topic will find this a user-friendly guide to the principles and practice of effective experimentation, the issues raised and the extensive references given will also be of interest to experienced researchers. My hope is that this book will offer the reader many opportunities for discovery and for meaningful research.

I would like to thank Henry Marcucella, Raymond C. Russ, Ellie McGowan, Brian H. Smith, and Ethel Tobach for their comments on earlier versions of the book and to Anna Yuan Miller and Daniela Brunner for their excellent renderings on which the figures were based. Thanks are due also to Gene Olson, who provided a slide of his crayfish operant device and a description of his technique. The help of Denis C. Gaffney of the library of the State University of New York—Health Sciences Center at Brooklyn is acknowledged. I would especially like to express my appreciation to Zhanna Shuranova of the Institute of Higher Nervous Activity and Neurophysiology in Moscow for many stimulating discussions on invertebrate learning and behavior. I would like to thank Valerie Montenegro, technical/production editor at APA Books, for once again making my life as a writer a little easier. I am grateful also to Judy Nemes, development editor at APA Books, for her patience, discernment, and skill in helping shape the book into its present form.

Finally, I invite readers who would like to share their comments regarding this book or the field of invertebrate learning to contact me at Oklahoma State University, Department of Psychology, North Murray 215, Stillwater, OK 74078-0250.

Introduction

Preview Questions

- What is learning?
- Why should we be interested in the learning of invertebrates?
- What are the distinguishing characteristics of invertebrates?
- How is this book organized?
- What are some milestones in the history of the use of invertebrates?

The Study of Learning

How does learning take place? Can people become more efficient learners and retain more of what they learn? Does learning exist in nonhuman species? How does a bee learn to go to a particular flower? Can the study of the nervous system of a squid reveal anything about the human nervous system? Is there a distinction between what people know and how they perform? How is learning defined? Answers to these and many other questions about learning are important from both a theoretical and a practical perspective.

As a student, for instance, you have a vested interest in finding more efficient study habits. As a teacher, I have an interest in improving my classroom technique and, based on my students' course evaluations, probably my clarity of expression as well. A parent has an interest in finding the best method of training his or her child. The neuroscientist has an interest in understanding the chemical changes associated with learning and memory in the hope of combating diseases that diminish the ability to learn and remember. The entomologist has an interest in studying the learning of insects in an effort to control the behavior of an insect pest. It should be clear from these examples that the study of learning is an enterprise shared by many scientific disciplines. As we will see throughout this book, the study of learning is enriched not only by the work of psychologists but also that of physiologists, biochemists, neuroscientists, biologists, zoologists, and entomologists.

A prerequisite for the study of learning—whether it is the learning of a family member, pet, insect, or amoeba—is that *learning* be clearly defined. This is not as easy as it sounds. As a mental exercise, take a few moments to consider the problem. Should learning, for instance, be linked with performance? Clearly not. How many times have you complained to your professor and parents that the exam you took did not reveal all that you have learned? Perhaps the questions were ambiguous, you were not motivated, or you suffered from test anxiety. Will your definition suggest that all behavior is learned? This suggestion also creates problems, for if we were to agree, we would be placed in the awkward position of having to include transient or short-term changes in performance associated with muscle fatigue or sensory adaptation as representing learning. How, then, should a definition of learning be framed?

Generally speaking, one can say that learning has occurred when conditions in the environment create a fundamental change in the organism so that there is a long-lasting change in the ability of the organism to respond. It should be noted that psychologists use the term *learning* more broadly than it is found in everyday usage. As Hill (1977) pointed out, learned behavior does not need to be "correct." For many, it is just as easy to learn bad habits as it is to learn good habits. In addition, learned behavior need not be conscious or deliberate. Perhaps the greatest contribution of a coach or teacher is pointing out the unwitting errors in reasoning or form. Moreover, attitudes and emotions can be learned in the same fashion as skilled movements. In other words, learning need not involve any overt behavior. More precisely, learning is defined as a relatively permanent change in behavior potential as a result of experience.

The latter definition contains several important principles. First, it must be noted that learning is inferred from behavior. One never observes learning directly. Rather, the process identified as learning is implied from observable data. Second, learning is the result of experience. This excludes changes in behavior produced as the result of physical development, aging, fatigue, adaptation, or circadian rhythms. Third, temporary fluctuations are not considered learning. Rather, the change in behavior identified as learned must persist as long as such behavior is appropriate. A fourth principle found in the definition is that more often than not, some experience with a situation is required for learning to occur.

Why Is Invertebrate Learning Important?

For many, the study of invertebrate learning is often the natural result of curiosity to understand and explore the natural world. I can still recall being amazed as a child at the intellectual achievement of Robbie, my pet cockroach, and years later—after I inadvertently infested the animal laboratories of Boston University with a particularly ugly strain of roach— standing in awe at the sheer number of cockroaches being driven out of their nests by the exterminator.

Consider for a moment how one's curiosity is so readily seized by some of the more common invertebrates. How many people, for example, wonder where the light of a firefly comes from? The honeybee is intriguing because it has a social structure that is not too different from that of human society, and there is, of course, that hairy, eight-legged monster that engenders in some a fear so irrational that psychologists have a special name for it: arachnophobia.

If the results of intrinsic human interest are to stimulate more investigation, then research must do more than titillate; it must lead to general principles. For those who seek basic principles of learning and memory, invertebrates offer several important advantages. For example, the varieties of invertebrate nervous systems created during evolution provide an excellent opportunity to begin to understand the human species. When the workings of the single cell of a protozoan, embryonic development in the sea urchin, the transmission of a nerve impulse traveling down the length of a squid giant axon, the visual system of the horseshoe crab, the escape system of a crayfish, the genetic dissection of behavior in the fruit fly, or the social behavior of the honeybee are understood, then science is that much closer to unraveling the mysteries of the human brain and human behavior.

In addition to what they can reveal about humankind, invertebrates are useful to test the generality of both behavioral theories of learning and any proposed underlying physiological or biochemical mechanisms. Invertebrate nervous systems are, on the whole, more amenable to physiological and biochemical manipulations than vertebrate nervous systems. Moreover, because the behavior of invertebrates is generally less complex and more reflexive than human behavior, the genetic analysis of physiological mechanisms is readily amenable to experimental manipulations. The fruit fly, *Drosophila*, and the nematode, *Caenorhabditis elegans*, have become the invertebrates of choice in the effort to identify genetic components in behavioral traits. If studies show, for instance, that a particular type of learning in a vertebrate depends on the presence of a cerebral cortex, a specific type of neural organization, or a certain level of cognitive development, then an invertebrate can reveal whether such features are always required. It may be found, for instance, that an invertebrate can learn the same task as a vertebrate but the underlying mechanisms or solutions are different. Invertebrates are also used to answer questions such as: When does learning first appear in the animal kingdom? Is it possible to trace the evolutionary development of a particular type of learning? Do invertebrates and vertebrates learn the same things, in the same ways?

The intriguing nature of invertebrate behavior, the accessibility of the nervous system, and the use of genetic manipulations combine to make invertebrates an attractive group to study in order to answer questions about learning and memory. This attractiveness is multiplied by the sheer number and diversity of invertebrates because it becomes easy to select behavior and physiology uniquely suited for a particular experimental design. For instance, if one is interested in the effect of regeneration of the nervous system on retention of a learned response, earthworms and planarians make fine subjects. Furthermore, if one is interested in how conditions inside a cell influence learning, one approach is to use aneural single cells of such protozoans as *Stentor* and *Paramecium*.

The Distinguishing Characteristics of Invertebrates

Before we examine the learning of invertebrates, it is important for you to have some idea of how these animals differ from vertebrates. One of the most identifiable characteristics of many well-known animals such as birds, lions, tigers, fishes, crocodiles, apes, and humans is the presence of a series of individual bones called *vertebrae*. In 1797, the biologist Jean Baptiste Lamarck (1744–1829) seized upon this characteristic as a basis for classifying the animal kingdom. Animals that possessed a backbone or vertebral column were called *vertebrates* by Lamarck, and those that did not possess such an anatomical feature became known as *invertebrates*.

In addition to the absence of a backbone, invertebrates can be distinguished from vertebrates by a variety of other features, such as their large numbers. About 97%, or 1,070,000 of all animal species are classified as invertebrates, with the remaining 10,000–20,000 species consisting of vertebrates. Another interesting fact is that invertebrates have inhabited the earth for a far greater period of time than have vertebrates. Fossil remains suggest that representatives of most invertebrate groups were established over 530 million years ago. As a point of reference, consider that the human species may have arisen as recently as 2 million years ago! Moreover, compared with vertebrates, invertebrates have successfully invaded every conceivable ecological niche from the terrestrial to the aquatic. They also come in a bewildering array of shapes and sizes. Their nervous systems range from the simple nerve nets of hydra to the complex ganglionic masses of such advanced arthropods as the honeybee. Anatomically, they are distinguished from vertebrates by the following features (Gardiner, 1972):

- Absence of a notocord, vertebral column, and internal bony skeleton
- Absence of an internal dorsal nerve cord
- Dorsal situation of the heart (if present)
- Development of specialized respiratory areas (if present) from the body wall, not the digestive tract

In addition to differences from vertebrates in anatomy, numbers, age, and extent of exploitation of the environment, there are significant differences in the organization of their nervous systems. Descriptions of some of the major differences follow (Bullock, Orkand, & Grinnell 1977; Corning, Dyal, & Lahue, 1976; Rosenzweig & Leiman, 1989):

1. **Morphology**. In invertebrates, the most common type of nerve cell is the monopolar neuron; the cell body of many invertebrate neurons is separated from the dendrites and is located on the "rind," or periphery, of the ganglion. Much of the integration of excitatory and inhibitory events occurs in the peripheral nervous system rather than in the central nervous system.
2. **Histology**. Much of the nervous system of invertebrates is replaced during metamorphosis; invertebrate ganglia typically contain large, repeatedly identifiable neurons and large axons. In contrast to those of vertebrates, the dendrites of neurons in in-

vertebrates are less clearly identifiable. Invertebrates possess far fewer nerve cells than vertebrates, and there is a simpler neural organization.

3. **Myelin sheath**. The tightly wrapped, fatty insulation surrounding an axon known as myelin is generally restricted to vertebrates. The exceptions appear to be the giant nerve fibers of earthworms and shrimps.

4. **Electrophysiology**. Spontaneous electrical activity of higher central ganglia in invertebrates are generally characterized by a wider range of frequencies than those found in vertebrates.

It should be mentioned that despite these differences, there are also similarities between the invertebrate and vertebrate nervous systems. These include mechanisms involved in the transmission of graded and action potentials, the opening and closing of tiny pores in the cell membrane known as ion channels, and the operation of biochemical substances known as neurotransmitters used to stimulate other neurons.

How This Book Is Organized

A Primer of Invertebrate Learning: The Behavioral Perspective is an introduction to the area of invertebrate learning. In this book I have tried to bring together some of the issues and present a few "tips" that must be considered before conducting a behavioral experiment using invertebrates. The book should be useful for those interested in acquiring a working knowledge of the behavioral techniques, data, and issues in the area of invertebrate learning. It is designed primarily as a supplemental text for courses in animal behavior and physiological psychology. It is especially useful as a supplement to my earlier book, *Invertebrate Learning: A Laboratory Manual and Source Book*, and provides additional source material.

The structure of this book is such that if you already possess some background information on, for instance, the design of experiments, you can proceed directly to those chapters that interest you most. If, for example, you would like to see a sample of the behavioral apparatus used in invertebrate research or read about some of the critical issues in the analysis of learning, by all means proceed directly to chapters 3 and 8! On the other hand, if you would like to learn something about the type of strategies used in invertebrate research and how to conduct learning experiments on, for example, habituation and classical conditioning, it would be best to start at the beginning.

The *Primer* consists of eight chapters, two appendixes, and a glossary. Each chapter concludes with a brief summary and several discussion questions. Here, in the Introduction, you have already learned about the importance of studying invertebrates and have discovered something about the major anatomical and physiological differences between vertebrates and invertebrates. Yet to come is a time line of 50 interesting facts related to the use of invertebrates in everyday life and in research.

Chapter 2 provides an outline of procedures for conducting an experiment on learning in invertebrates. In that chapter you will see, and learn how to select from, the major approaches that are available for the study of invertebrate learning. These include the comparative psychology approach and the simple system approach to invertebrate learning. In addition, preliminary information is provided on the use of basic learning paradigms, apparatus, and the importance of the comparative method. A unique feature of chapter 2 is a discussion of some of the important skills a good invertebrate researcher must possess, and I have included a handy form for planning and reporting experiments. This form will help you not only design your experiments but also examine the results of others.

Chapter 3 presents—for the first time in a single source—many illustrations of the type of apparatus used in invertebrate learning research. In addition, tips are provided on how to use them and how to select a particular apparatus to suit a particular research question.

Chapters 4, 5, and 6 present some of the research findings in the areas of nonassociative and associative learning, respectively. Habituation and sensitization are covered in chapter 4, classical conditioning in chapter 5, and instrumental and operant conditioning in chapter 6. Each chapter contains handy tables that will enable you to identify quickly not only the invertebrate used in such learning procedures but also what stimuli have been used. They also contain some important discussions on the type of control procedures that should be used.

Chapter 7 provides an overview of the cellular mechanisms that have been found to underlie the learning of invertebrates. The treatment of the cellular mechanisms is necessarily brief because to understand them properly, one must have a good background in physiology (biochemistry and neuroanatomy are useful, too!). The greater value in chapter 7 is that it discusses the strategies involved when undertaking a search for the cellular analysis of learning.

The final chapter, chapter 8, is a discussion of several behavioral issues that I believe are central to the analysis and interpretation of invertebrate learning experiments. Many of these issues are no longer discussed in the literature of invertebrate learning, yet they remain important, and you should be made aware of them. Many of the issues are somewhat technical, however, and I advise you to read the other chapters first.

The remaining sections of the book consist of a glossary and two appendixes. The first appendix contains tables of source material such as review articles, apparatus, and general interest books. The second appendix presents the names and addresses of some investigators in the area of invertebrate learning. This information will enable you to contact scientists whose work excites you. The glossary contains the definitions of terms you are likely to encounter in your readings.

To better acquaint you with the many contributions invertebrates have made to science, I will close this chapter with a chronology of 50 "great moments" in the use of invertebrates, an "invertebrate hit parade" to guide you through some of the scientific achievements in history made possible by the "cooperation" between scientist and invertebrate.

This list is not meant to be inclusive. Rather, its purpose is to show that the study of invertebrates is a vibrant and vital area of research that affects a surprising amount of our daily activity. Any compilation of this type is of course subjective, and an attempt was made to include facts not generally known. For instance, the founders of classical conditioning, operant conditioning, and psychoanalysis early in their careers all studied invertebrates. It is probable that these invertebrate studies represent the first publications of I. P. Pavlov, B. F. Skinner, and S. Freud. In addition, discoveries made using invertebrates have led directly to 10 Nobel Prizes in Physiology or Medicine, one Pulitzer Prize, the first unique contribution of an American biologist to science, and the first laboratory study of animal behavior.

50 Milestones in Invertebrate History

🐝 7000 B.C. The earliest known honey harvesting by humans took place. For an entertaining account of the domestication of bees in the ancient world, see Crane and Graham (1985).

- 1744. Abraham Trembley founded experimental zoology. Using hydras—or as they were originally known, polyps—he was the first to discover and describe the ability of animals to regenerate. When he sliced a hydra into as many as eight pieces, he was astonished—as were others in his day—to find that each piece regenerated into a complete animal. This demonstration was one of the first challenges to the widely held belief that mating was required for reproduction. Moreover, as one would imagine, the results caused quite a stir among theologians who were concerned about the soul of such a regenerated animal. In addition to his work on regeneration, he invented—again using hydra—the grafting technique in which part of one animal is transplanted onto a second animal. In 1744, Trembley published his research on hydra in a book entitled *Mémoires, Pour Servir à L'histoire d'un Genre de Polypes D'eau Douce, à Bras en Forme de Cornes* (Memoirs concerning the natural history of a type of freshwater polyp with arms shaped like horns). Of his many contributions, perhaps the most important was that he stimulated the experimental analysis of invertebrate behavior. Those interested in learning more about Trembley can consult Baker (1952) and Lenhoff and Lenhoff (1988).

- 1793. Jean Lamarck (1744–1829) was appointed professor of "inferior animals" at the National Museum of Natural History in France. He promptly coined the term *invertebrate*. See Mayr (1982) for a discussion of Lamarck's contribution to invertebrate behavior.

- 1872. The Zoological Station of Naples was founded. For many, the Stazione Zoologica is considered the first great marine biological station. Florey (1985) entertainingly covered the role that the Stazione played in revealing the secrets of the neuron.

- 1878. Charles Whitman provided a detailed description of the development of sea urchin embryos. His results and theoretical insights gained from the study of sea urchins are generally considered the first unique contribution of an American biologist to science.

- 1880. Thomas Huxley published *The Crayfish: An Introduction to the Study of Zoology*. This book was one of the first to demonstrate that invertebrates can be used to study behavior.

- 1882. Sigmund Freud published his investigation on crayfish nerve fibers and nerve cells. In it, he demonstrated that the axon is an extension of the cell body (Freud, 1882). While working in the physiology department, Freud met Joseph Breuer. Freud's contribution to neurobiology has been discussed by Florey (1990).

🕮 1885. Ivan Pavlov published a paper on shell opening in clams (Pavlov, 1885/1951). Throughout his career, Pavlov extolled the virtues of the study of invertebrates.

🕮 1887. George W. and Elizabeth G. Peckham published an article describing habituation to vibration in spiders. Their work has the distinction of being generally recognized as the first experimental study of animal learning (Peckham & Peckham, 1887).

🕮 1888. The Marine Biological Laboratory (MBL) in Woods Hole, Massachusetts was founded. For a history of the MBL, see Lillie (1944/1988).

🕮 1892. Charles Turner published his first paper in psychology, entitled *Psychological Notes Upon the Gallery Spider: Illustrations of Intelligent Variations in the Construction of the Web* (Turner, 1892a). His contributions to the study of invertebrate learning were many and varied. He is perhaps best known as the developer of the choice-chamber method to study cockroach behavior (Turner, 1912). This method, which is still used today, requires a roach to inhibit its tendency to enter a dark compartment by pairing each entry with an electric shock. What is perhaps not so well known is that he was the first psychologist of African-American descent and possibly the first African American to publish in *Science* (Turner, 1892b). For a full account of this fascinating scientist, I would urge you to consult Cadwallader (1984).

🕮 1894. Vladimir Wagner (1894) published a monograph entitled *Industry of Spiders*, in which he demonstrated the importance of comparative methods of behavior analysis. Wagner's behavioral techniques and theoretical insights often anticipated similar developments in the West. Wagner is credited with being the founder of comparative psychology in Russia.

🕮 1901. Adele Fielde adapted William Small's newly developed maze technique (developed for rats) to studies of ant learning.

🕮 1902. Ronald Ross received the Nobel Prize for Physiology or Medicine for identifying the *Anopheles* mosquito as the transmitter of malaria.

🕮 1906. Herbert Jennings (1906/1976) published experiments demonstrating habituation to touch in protozoans. Jennings persuasively argued that such adaptive behavior could not adequately be explained by simple trophic reactions of the sort proposed by Jacques

Loeb, and therefore he opened the door for an evolutionary analysis of cognition, which continues to the present day.

- 1907. Charles Laveran received the Nobel Prize for Physiology or Medicine for research on diseases caused by protozoans.

- 1914. Axel Melander became the first person to document the resistance of insect pests to insecticides.

- 1929. Theodore Schneirla published his dissertation describing the learning and orientation of ants. This study marks the beginning of the career of one the giants of comparative psychology in America. In addition to Schneirla's well-known studies of ant learning, some of his more notable contributions include the classic textbook *Principles of Animal Psychology*, written with Norman Maier (Maier & Schneirla, 1935/1964); his study of migratory behavior of army ants; and his theoretical insights into the development and evolution of behavior. A fine selection of his writings is available in Aronson, Tobach, Rosenblatt, and Lehrman (1972).

- 1930. Burrhus F. Skinner published his first scientific paper. The topic was an analysis of the geotropic response in the ant (Skinner & Barnes, 1930).

- 1933. Thomas Morgan received the Nobel Prize for Physiology or Medicine for discovering the role of chromosomes in heredity. His animal of choice was the fruit fly, *Drosophila*. He was also the first American-born nonphysician to receive the prize in Physiology or Medicine.

- 1934. C. Ladd Prosser (1934) and J. H. Welsh (1934) published their results on the behavior and physiology of one of the first identified neurons—the crayfish caudal photoreceptor. The historical significance of this work has been discussed by Wilkens (1988).

- 1936. John Z. Young (1936) published his work on the nervous system of cephalopods. He suggested that the giant axons of the squid—which he discovered may be as much as a millimeter in diameter—would provide excellent material for research on the propagation of nerve impulses.

- 1938. Ralph Buchsbaum published *Animals without Backbones*, a thorough and entertaining introduction to the major invertebrate animals. This book, in various editions, has been in use for more than 50 years. In addition to its long "shelf life"—it was last revised in 1987—the book has the distinction of being the first biology textbook to be reviewed by *Time* magazine.

📖 1940. Gottfried Fraenkel and Donald Gunn published *The Orientation of Animals* (1940/1961). It is generally considered the most complete summary of directed movements of invertebrates, and it contributed to the lexicon of behavior an objective language to describe the movement of animals. Much of the book is devoted to the behavior of invertebrates and is a fine example of how concepts designed to explain invertebrate behavior can contribute to an understanding of vertebrate behavior.

📖 1940. Libbie Hyman (1940–1967) published the first volume of the monumental work *The Invertebrates*. She wrote five additional volumes, the last appearing in 1967, when she was 78 years old.

📖 1944. H. Frings published a paper on classical conditioning of proboscis extension in insects. Although his primary interest was in developing a technique for measuring thresholds, olfactory conditioning has become arguably the most powerful and widely used conditioning procedure for measuring learning in invertebrates.

📖 1946. Hermann Muller received the Nobel Prize for Physiology or Medicine for discovering that mutations can be created by X-ray irradiation. The study of mutations and the development of techniques to create them rapidly were carried out on the fruit fly.

📖 1948. Paul Müller received the Nobel Prize for Physiology or Medicine for demonstrating the insecticidal properties of dichloro-diphenyl-trichloro-ethane, better known as DDT. The long-term deleterious effects of DDT on both the environment and food chain were not known then but have since led to its disuse in the United States.

📖 1949. Kenneth Cole (1949, 1968) published a method for measuring membrane currents. The method, known as voltage clamp, was developed using the giant axons of the squid.

📖 1950. Commencement of the Pavlovian Session in Physiology of the All-Union Academy of Sciences and the Academy of Medical Sciences of the U.S.S.R. took place. The leading participants of this conference, who were Communist functionaries, effectively eliminated the use of invertebrates as subjects for comparative investigation. This was accomplished by, for example, restricting funding for invertebrate work and purging from the academic ranks those scientists whose research plan did not fit with the Stalinist ideal. The effect of this purge—although lessened with the passage of time—is still felt today. Those interested in the history of psychology in Russia should see Joravsky (1989).

🐛 1955. Two graduate students, Robert Thompson and James McConnell (1955), published a paper suggesting that planarians can form an association between increase in illumination and shock. The demonstration of classical conditioning in planarians spawned a generation of "worm runners." At the time, the planarian experiments generated considerable interest, and the results of such experiments contributed greatly to knowledge of the underlying mechanisms of learning and memory. Some of the more controversial issues revolved around whether a light-conditioned stimulus can be considered valid for a planarian because it elicits—prior to pairing—behavior that resembles the conditioned response. The problem of alpha conditioning was brought into sharp focus by the planarian experiments and remains unresolved today. Controversy erupted several years later, when it was subsequently shown that memory of the conditioned response not only can survive regeneration but also can be transferred to untrained animals when they are permitted to feed on trained "donors" (McConnell, 1962; McConnell, Jacobson, & Kimble, 1959). A review of these experiments can be found in Corning and Kelly (1973).

🐛 1959. Stanley Ratner and Kliem Miller demonstrated that dark-adapted earthworms can be classically conditioned by pairing vibration with illumination. This simple and elegant technique was the first to conclusively demonstrate classical conditioning in the earthworm and served to popularize the use of the earthworm as an animal to study the physiological and biochemical basis of learning and memory.

🐛 1960. The first volume of *The Physiology of Crustacea*, edited by Talbot Waterman (1960, 1961) was published; Volume 2 followed in 1961. These volumes mark the first time that a comprehensive review of crustacean physiology appeared in English.

🐛 1962. Adrian Horridge published an experiment in which a headless cockroach learned to keep one leg raised to terminate shock. This experiment was one of the first to suggest that invertebrates can be used to explore questions about the neuronal basis of learning and memory. Leg position learning—or the Horridge paradigm, as it has become known—is used to demonstrate learning in a wide variety of situations ranging from intact animals to single ganglia.

🐛 1962. Rachel Carson published *Silent Spring*, which described in vivid detail the environmental disaster brought about by the unrestricted use of insecticides and pesticides. Its publication and wide readership was recognized as a catalyst for the environmental protection movement; it is still good reading. Those interested in Rachel Carson and the controversy surrounding publication of *Silent Spring* should see Graham (1970).

1963. Alan Hodgkin and Andrew Huxley shared (with John Eccles) the Nobel Prize for Physiology or Medicine for discoveries relating to the generation of the nerve impulse. Among their technical achievements was the development in 1939 of a technique for inserting a microelectrode into the giant axon of the squid, which made possible direct intracellular measurements of membrane potentials. (In the same year, H. J. Curtis and K. S. Cole developed a similar procedure.) However, World War II delayed publication until 1945 (Hodgkin & Huxley, 1945).

1965. Eric Kandel and Ladislav Tauc (1965) studied neuronal analogues of classical conditioning in *Aplysia*. Their success in exploiting the *Aplysia* nervous system for studies of learning and memory helped to establish and stimulate the growth of the "simple systems" approach to behavior analysis. Subsequent research established *Aplysia* as one of the most important sources of information about the neuronal mechanisms of learning.

1965. Theodore Bullock and Adrian Horridge's two-volume treatise *Structure and Function in the Nervous Systems of Invertebrates* was published. These volumes stand as the classic guide to anatomical and histological data on invertebrates.

1967. Seymour Benzer introduced the "countercurrent method" for genetic dissection of behavior in the fruit fly, *Drosophila*. For many years, this was the technique of choice for the genetic analysis of learning and memory, and variations are still generating important data today. For a discussion of the technique, see Benzer (1973).

1967. H. Keffer Hartline shared the 1967 Nobel Prize in Physiology or Medicine (with George Wald and Ragnar Granit) for his discoveries concerning the physiological and chemical processes in the visual system. Many of his insights were gained from studying the optic nerve of the horseshoe crab, *Limulus*.

1968. E. Kravitz and A. Stretton introduced the use of Procion dye to stain individual living nerve cells of the lobster abdominal ganglion. The Procion method soon became the intracellular staining technique of choice for a wide variety of invertebrate and vertebrate nervous systems. A historical introduction to the Procion method is available in Stretton and Kravitz (1973).

🐚 1969. Franklin Krasne demonstrated habituation of the tail-flip escape response of crayfish. Access to the lateral giant fiber controlling the escape reflex has made the crayfish a popular animal for simple systems research.

🐚 1970. Bernhard Katz shared the Nobel Prize in Physiology or Medicine (with Ulf von Euler and Julius Axelrod) for discoveries concerning nerve transmission. As is so often the case, much of the work on nerve transmission was performed on the squid giant axon.

🐚 1973. Karl von Frisch shared the Nobel Prize (with Konrad Lorenz and Nikolaas Tinbergen) in Physiology or Medicine for his work on the behavior and "language" of honeybees. One of his more significant contributions was the development of the free-flying technique, in which free-flying foragers are trained to come from the hive to the laboratory.

🐚 1973. The first in a three-volume set entitled *Invertebrate Learning* by William Corning, John Dyal, and Dennis Willows (1973–1975) was published. This represented the first comprehensive review of invertebrate learning since Volume 2 of Warden, Jenkins and Warner's, *Comparative Psychology* (1940)—a span of 33 years.

🐚 1973. George Mpitsos and William Davis demonstrated taste aversion learning in the marine mollusc *Pleurobranchaea*. The ease of conditioning and the accessibility of the molluscan nervous system continue to make this animal one of the most popular in simple system research.

🐚 1974. Daniel Alkon (1974) demonstrated the suppression of phototaxis by associative learning in the nudibranch mollusc *Hermissenda*. *Hermissenda* subsequently became one of the most important sources of information about the neuronal basis of learning and memory. For a description of the use of *Hermissenda*, see Alkon (1983).

🐚 1975. Alan Gelperin demonstrated taste aversion learning in the garden slug, *Limax maximus*. Like molluscs and crustaceans, *Limax* becomes a popular animal for simple system research.

🐚 1991. Bert Hölldobler and Edward Wilson (1990) won the Pulitzer Prize for general nonfiction for *The Ants*, an exhaustive review of what is known about the anatomy, physiology, social organization, ecology, and natural history of ants.

🔖 1991. Erwin Neher and Bert Sakmann shared the Nobel Prize in Physiology or Medicine for developing the patch clamp technique used to reveal the function of single ion channels. Hundreds of laboratories have used the patch clamp technique since its use on denervated frog muscle fibers in 1976. Prior to 1976, however, the patch clamp technique was used on the soma of neurons in the snail *Helix pomatia* (Neher & Lux, 1969). An account of the development and uses of the technique can be found in Neher and Sakmann (1992).

Summary

In this section, I have discussed why the study of learning is important and, in particular, how invertebrates can shed some light on the learning process. We have seen how invertebrates differ both anatomically and physiologically from vertebrates. In addition, we have seen how the study of invertebrates is a vibrant and vital area of science that affects a surprising amount of our daily activity.

Discussion Questions

- What are some similarities and differences between invertebrates and vertebrates?
- How do invertebrates influence your daily activity?
- What are some advantages that invertebrates provide for the study of learning?
- Why would you be interested in the study of invertebrate learning?
- Is the learning process different for vertebrates and invertebrates?

2 Constructing an Experiment With Invertebrates

Preview Questions

- What are the major strategies for the study of invertebrates?
- How is a particular research strategy chosen?
- What are some important skills for those planning to conduct invertebrate research?
- What are the major varieties of learning?
- What is the comparative method?
- How is the comparative method used in the study of invertebrate learning?

Major Strategies Used in the Study of Invertebrate Learning

In this chapter, we will take a brief tour of four major strategies developed to analyze invertebrate learning. The first strategy we will examine is that taken by the comparative psychologist. We will then move on to the ethological approach, the behavioral genetics approach, and the simple systems approach. As we progress with our discussion, you will notice that most of the strategies have common features.

Such a close familial relationship is to be expected because each strategy is designed to find and analyze similarities and differences in the hope of developing a more complete picture of some anatomical feature, neural mechanism, or behavior. The search for similarities and differences is known as the comparative method and is the common ancestor, or if you prefer, the glue that binds these strategies together. The distinction between strategies is purely arbitrary—a fact that you will experience as you progress through the study of invertebrate learning.

The Comparative Strategy

Comparative psychology is loosely defined as the study of similarities and differences in the behavior of animals. The principal impetus for the creation of comparative psychology arose from the evolutionary theory Charles Darwin put forth in *The Origin of Species* (1859/1936) and subsequently reinforced in his *Expression of the Emotions in Man and Animals* (1872/1965). Darwin believed that mental or cognitive functions are shaped by the same pressures that give rise to the anatomical features of a species—that there is no property of mind that is not present in an incipient form in some lower animal. The data on which he based this idea—exemplified by the work of George Romanes, a colleague of Darwin's—consisted of anecdotal material gathered by observing the behavior of animals such as cats, dogs, and ants. Criticisms of the anecdotal method from such behavioral scientists as Sir John Lubbock, C. Lloyd Morgan, Jacques Loeb, and Vladimir Wagner led directly to the application of the experimental method to the comparative analysis of behavior. Thus, rather than catalog anecdotes, or stories about learning, scientists began to actively manipulate what they thought were the conditions that produced learning.

Initially, the purpose of the early comparative studies of learning was to establish that animals, particularly invertebrates, could be trained. The goal of such studies was rather simple: to discover the presence or absence of learning ability. For example, H. S. Jennings (1906/1976), working in the tradition of the "mental" phylogenic scale established by Darwin and Romanes, argued persuasively that habituation of protozoans to touch could not be explained by the simple trophic reactions proposed by Loeb (for summary, see 1918/1973), and therefore protozoans cannot be automatons.

Such experiments ushered in the "golden age" of comparative psychology, in which the behaviors of many animals—both invertebrate and vertebrate—were studied under field and laboratory conditions and various theories were advanced to account for them. Around the turn of the century, many techniques were developed to study learning. Some of the more memorable techniques of the period include E. L. Thorndike's (1911) puzzle box, in which he investigated the learning of various vertebrates such as chickens and cats, W. S. Small's (1901) adaptation of the Hampton Court maze to the study of learning in rats, and C. H. Turner's (1914) conditioning of moths by pairing vibration with electric shock.

The study of invertebrate behavior was particularly active during this period. A count of all papers appearing in the *Journal of Animal Behavior* (the forerunner to the *Journal of Comparative Psychology*) in the years 1911 through 1915 reveals that approximately 40% of all the studies reported employed invertebrates (Beach, 1950). Unfortunately, as the years passed, the number of papers devoted to the "psychological" study of invertebrates declined to less than 5% and was often zero (Bitterman, 1960).[1]

In spite of the low number of articles devoted to invertebrate learning, the objectives of comparative psychological analysis of invertebrate behavior remain the same as they were when Fielde (1901) adapted Small's rat maze to the study of ant behavior at the turn of the century. As Waters (1960) pointed out in discussing the role of comparative psychology, the basic strategy is to:

1. Identify and classify the behavioral repertoire of animals.
2. Establish relationships between the behaviors of animal species.
3. Trace the ontogenetic development of behavioral processes and social organization both within an individual and between phylogenic groups.
4. Characterize the similarities and differences in behavior across the evolutionary scale.
5. Create comprehensive theories that summarize and predict behavior across the evolutionary scale.

The methods employed by contemporary comparative psychologists are as rich and varied as the problems psychologists attempt to answer. They range from the techniques of behavioral genetics and cellular physiology to naturalistic observation. Unfortunately, the use of naturalistic observation as a vital method in comparative psychology has been overlooked by many scientists outside the area of comparative psychology. They sometimes have the mistaken impression that comparative psychologists do not study behavior under natural conditions but prefer to imprison the animals and force into the open the secrets of behavior.

---

1. Recently, Gallup (1989) plotted the number of papers appearing in the *Journal of Comparative Psychology* since its predecessor, the *Journal of Comparative and Physiological Psychology*, was split into the *Journal of Comparative Psychology* and *Behavioral Neuroscience*. The percentage of total papers using invertebrates for the years 1983 through 1987 is disappointingly low, the percentages being 8%, 5%, 12%, 5%, and 2%. I have calculated the percentages for the years 1988 through 1991, and there is no change in the pattern: 11%, 4%, 6%, 6%. The situation in *Behavioral Neuroscience* is no better. I have calculated the percentages of invertebrate papers for the years 1983 through 1991 and once again the percentage is low: 2%, 2%, 1%, 2%, 5%, 1%, 1%, 4%, and 3%, respectively.

This misconception has been so popularized over the years that it now serves to distinguish the disciplines of comparative psychology and ethology. Such a dichotomy, in my view, is misleading and encourages the erroneous belief that one discipline is more "correct" than another. Perhaps the best statement on the use of naturalistic observation in comparative psychology can be found in the works of T. C. Schneirla (1950).

The contribution that comparative psychology has made to the study of invertebrate learning is substantial and impressive. Material on the history of comparative psychology can be found in any textbook on comparative psychology. I would recommend, however, Warden, Jenkins, and Warner's (1935) text because it provides an authoritative review and provides photographs of not only a number of early contributors to the field but also some techniques of the period. The book by Richards (1987) provides an excellent historical introduction of how critical the study of invertebrate behavior was to the success of Darwin's theory of evolution.[2]

Reviews of techniques developed for the study of learning can be found in Watson (1914), Warden, Jenkins, and Warner (1935), Fraenkel and Gunn (1961), and Bitterman (1962). Chapter 3 in this volume provides numerous examples of apparatus developed for the study of invertebrate learning.

The Ethological Strategy

Ethology is generally defined as the study of animal behavior in the natural environment. Like comparative psychology, the creative force behind the development of ethology was Darwin's theory of evolution. This familial relationship has suggested to some that comparative psychology and ethology are "twins reared apart" (Gray, 1973). Like comparative psychology, ethology attempts to understand behavior from a number of perspectives, including ecological, functional, and physiological. As described by Harré and Lamb (1986), classical ethology had its origins in the work of 19th-century zoologists who were interested in the

---

2. For those interested in further reading, a discussion of the history of comparative psychology and its development in Latin America, Europe, and Russia is available in Tobach (1987). A historical introduction to the comparative analysis of learning is available from Bitterman (1967b). Applications of comparative analysis to the study of learning can be found in articles by Riesen (1960), Maier and Schneirla (1964), Lester (1973), Bitterman (1975, 1988), Thomas (1980), and Macphail (1987). Lester provides a particularly good discussion of the many difficulties encountered in creating phylogenic scales of learning. An updated discussion of this issue is provided by Campbell and Hodos (1991). Bitterman (1967a) provides an entertaining account of his early work in a *Scientific American* article.

observation of natural behavior and how such behavior enabled the animal to survive and reproduce: characteristics that were to become the hallmark of ethology. Because of this kinship to biology and evolutionary theory, ethology is considered a branch of biology.

Early contributors to (what was to become) ethology include the American zoologist Charles Whitman, who maintained that behavior and anatomy should be studied within an evolutionary perspective, and the German zoologist Jakob von Uexküll, who suggested that an instance of behavior can only be understood in the context of the entire behavioral repertoire of the animal. Thomas Huxley combined observational technique with an evolutionary perspective in his seminal work *The Crayfish: An Introduction to Experimental Zoology* (1880/1973). For historical overviews of ethology, see Thorpe (1979), Hinde (1982), Hansell (1985), or Baerends (1988). An informative article describing the historical antecedents that precipitated the split between ethology and comparative psychology is available from Jaynes (1969).

Since the work of Konrad Lorenz and Niko Tinbergen in the 1950s, there has been a tremendous increase in the use of the ethological strategy in the analysis of behavior. The use of naturalistic observation, the range of behavior that is explored, the importance that is placed on describing the functional significance of behavior, and the rich theoretical concepts that have been developed to explain behavior (such as the *fixed action pattern*, sign stimuli, releasers, and supernormal stimuli) are as opulent as any Freudian description of the mind and have permitted ethology to integrate itself successfully with the research areas of ecology, genetics, neurophysiology, and sociobiology.

As Tinbergen (1951) described in *The Study of Instinct,* the basic strategy of ethology is to divide the analysis of behavior into four steps:

1. To identify the developmental sequences of behavior.
2. To relate the behavior to an underlying physiological mechanism(s).
3. To uncover the adaptive and functional significance of behavior.
4. To examine behavioral change through evolutionary time.

The methods employed by ethologists are as varied as those used by comparative psychologists and include laboratory and field investigations as well as the sophisticated instruments of the neuroscientist. Of special interest, however, are three techniques uniquely associated with ethology: the ethogram, the deprivation experiment, and the use of models.

The *ethogram* was developed from the anecdotal material gathered by George Romanes in the mid-nineteenth century. The ethogram employs observational techniques, often under natural conditions, to create a catalog of the basic repertoire of a species. These catalogs are invaluable in their own right and serve as a guide to those interested in neuroethology and comparative psychology. Some well-known ethograms are found in the writings of Karl von Frisch, Jane Goodall, and Dian Fossey. Their work provides invaluable information on many facets of the behavior of honeybees, chimpanzees, and gorillas, respectively.

An understanding of your invertebrate subject in its natural environment is of critical importance in the design and interpretation of your preparation, in the selection of an appropriate combination of stimuli and reinforcers, in the construction of apparatus, in the maintenance of your subjects, and in tracing the functional significance of behavior across evolutionary time. Moreover, it allows you to observe behavior that is not easily captured in the laboratory and permits you the opportunity to generalize the laboratory findings. An ethogram is one way of obtaining background information on your invertebrate. It should be mentioned that because behavior is so complex, the construction of an ethogram is not a simple matter. For example, decisions must be made on which behaviors to include and exclude, how an instance of behavior is defined and recorded, and how your behavioral categories relate to those established by other scientists. Despite these concerns, ethograms and behavioral profiles are a important first step in the analysis of behavior. A discussion of the issues involved in constructing an ethogram can be found in Schleidt, Yakalis, Donnelly, and McGarry (1984), Drummond (1985), Gordon (1985), and Leonard and Lukowiak (1985).

A second technique associated with ethological investigation is the *deprivation*, or isolation experiment. Originally conceived by Douglas Spalding in the mid 1880s, the deprivation experiment allows ontogenic or developmental factors to operate on behavior in the absence of environmental influences. A common example is the rearing of animals in isolation with the aim of teasing apart genetic and environmental factors that influence social interaction. Such experiments led to the discovery that normal behavioral development occurs by the unfolding of genetically based tendencies. The deprivation experiment is not without problems, however, the most obvious being that deprivation may damage the animal. In addition, it cannot be assumed that behavior emitted following deprivation is innate. A deprived animal is still a functioning animal sensitive to environmental contingencies.

A third technique popularized by ethologists is the use of animal "mannequins" or *models*. Such models are extremely useful in determining, for instance, the critical features of stimuli necessary to elicit fixed action patterns. For a discussion of ethological techniques, see Tinbergen (1963) and Fantino and Logan (1979). A collection of *Scientific American* articles on psychobiology contains several papers by ethologists such as Lorenz and Tinbergen (McGaugh, Weinberger, & Whalen, 1967).

From the creation of ethograms of invertebrates known to be important for the cellular analysis of learning to the design of experiments demonstrating the importance of ecological variables, the contribution of ethology to the study of invertebrate learning has been great. Of special significance to those readers interested in the use of invertebrate nervous systems in the analysis of learning is the development of neuroethology, which is a hybrid of neurobiology and ethology. As Camhi (1984) has discussed, neuroethology attempts to understand the function of the nervous system underlying natural behavior. In the best cases, neuroethologists attempt to investigate the underlying physiological mechanism of behavior in freely moving animals. Invertebrates are often good model systems for neuroethologists because they are tough, have complex behavior that is accessible with neurobiological techniques, and are convenient to use.

Behavioral Genetics

Behavioral genetics is loosely defined as the study of how genes lead to the expression of behavior. Behavioral genetics is characterized by an exciting blend of Darwinian evolution, ecology, learning theory, and molecular biology.

As in the case of comparative psychology and ethology, the principal impetus for the development of behavioral genetics was the evolutionary theory of Charles Darwin. Francis Galton, Darwin's cousin, became a central figure in extending Darwinian theory to the inheritance of mental characteristics. Galton's contributions to behavioral genetics are many and varied and include the development of apparatus to measure mental abilities, the design of such statistical measures as correlation and percentile, an insistence on quantification and measurement, and the introduction of the twin-study technique to assess the role of inheritance and environment on behavior.

Much of the early research in behavioral genetics was directed at demonstrating that genes influence behavior. The use of invertebrates has traditionally been strong in behavioral genetics. Some of the early work includes an analysis of sexual behavior, behavior associated with movement toward a spot of light (i.e., phototactic), and behavior associated with the effect of gravity (i.e., geotaxic behavior) in the fruit fly, *Drosophila*. For a review of this work, see Hirsch and Erlenmeyer-Kimling (1967). Review of contemporary work on *Drosophila* is available in Tully (1991) and in the citations of Table A-1 in appendix A.

The relationship between learning and genetic factors has been recognized since the classic work of Tryon (1930), in which maze-"bright" and maze-"dull" rats were selectively bred. Invertebrates offer to those interested in behavioral–genetic analysis many advantages. For instance, the well-developed genetic methodology, small size, brief life span, ease of maintenance, prodigious breeding, and wide range of behavior have encouraged the use of the fruit fly, *Drosophila* and the nematode, *C. elegans*. One of the first demonstrations of learning in fruit flies conducted within a behavioral–genetic framework was performed by Quinn, Harris, and Benzer (1974). The Quinn et al. experiment was important because it presented a technique for rapid isolation of mutants with altered abilities to learn. These mutations can then be analyzed using a number of sophisticated biochemical, neuroethological, and genetic methodologies. Some of the more famous mutants include *Amnesiac*, *Dunce*, and *Rutabaga*.

The most comprehensive review of behavioral–genetic experiments conducted prior to 1940 can be found in Hall (1940). Excellent introductions to the area are available in Fuller and Thompson (1960, 1978) and McClearn and Foch (1988).

Fuller (1960) described the basic objectives of the behavioral–genetic strategy:

1. To determine if individual differences in behavior are inheritable.
2. To identify the genetic mechanisms producing behavioral variation.
3. To discover how genes influence behavior.
4. To assess the contribution of genetic and environmental factors in behavior.
5. To uncover how genetic variation within a species affects social organization.
6. To locate the chromosomal position of a gene or genes that influence behavior.

The methods employed by behavioral geneticists are a blend of behavioral techniques developed by comparative psychologists and ethologists and the powerful techniques developed by geneticists, neuroscientists, and molecular biologists. A well-known technique is the use of single-gene mutations to "dissect" behavior. According to this technique, mutations are created by such procedures as selective breeding and subjecting animals to X rays. An attempt is made to relate behavioral differences between mutants and normal animals in terms of the action of genes affecting physiology, neural structures, and neurochemistry. A related method is the strain comparison, in which the behaviors of different strains of animals of the same species are compared. An example of this method is the twin-study technique. Quantitative statistical methods have also been developed to search for complex traits, such as learning ability, within a population.

For a discussion of behavioral–genetic techniques, see *Behavioral Genetics: A Primer* by Plomin, Defries, and McClearn (1980). The book by Fuller and Thompson (1978) and the article by McClearn and Foch (1988) are also very informative. An entertaining *Scientific American* article on the use of *Drosophila* is available in Benzer (1973).

The contribution of behavioral genetics to the study of behavior and to learning in particular has been substantial. One of the most significant contributions was the discovery that both environmental and genetic factors interacted to influence behavior. Another important contribution was in the simple systems approach to behavior analysis and learning. The area of behavioral genetics has also contributed to the birth of sociobiology. Sociobiology is concerned with the biological basis of social behavior, and much of its methodology is taken from ethology and behavioral genetics.

The Simple Systems Strategy

The *simple systems strategy* is generally defined as the study of the cellular basis of behavior. The basic strategy is to identify the neural circuits associated with a particular behavior and attempt to reveal the cellular changes involved (Carew & Sahley, 1986). The simple systems approach represents one of the best examples of the effectiveness of an interdisciplinary approach to problem solving. Through the combined efforts of anatomists, biochemists, biologists, ethologists, entomologists, geneticists, naturalists, neuroethologists, psychologists, physiologists, and zoologists, scientists are rapidly approaching an understanding of the neuronal circuitry and cellular changes associated with behavioral plasticity.

The simple systems approach has been predominantly identified with the study of invertebrate learning and will be discussed in chapter

7. Some well-known examples of this approach include habituation, sensitization, classical conditioning, and operant conditioning of gill withdrawal in *Aplysia*; classical conditioning and conditioned suppression of phototaxis in *Hermissenda*; classical conditioning in bees; odor aversion learning in various molluscan species; escape behavior in crayfish; and operant conditioning in various insects and crustaceans. However, advances in the neurosciences have now made it possible once again to extend the simple systems strategy to the study of vertebrate learning. These include classical conditioning of the nictitating membrane in the rabbit, eye-blink conditioning in the cat, conditioning of isolated spinal reflexes in cats, cardiac conditioning in the pigeon, and olfactory conditioning in neonatal rats.

The simple systems strategy is considered to have started with the publication of Adrian Horridge's study of leg lift learning of intact and headless insects (1962). This study was one of the first—along with the early planarian classical conditioning experiments—to suggest that invertebrates can be used to explore the cellular changes associated with learning and thereby bypass some of the early difficulties in identifying cellular changes associated with vertebrate learning. The use of invertebrates as tools in the analysis of learning gained significant momentum following the publication of *Chemistry of Learning: Invertebrate Research*, edited by Corning and Ratner (1967). This volume contains numerous examples of the use of invertebrates in the analysis of learning and is still worth reading. The basic objectives of the simple systems strategy have been identified by Entingh et al., (1975):

1. To identify the neurochemical systems involved in the formation of memory.
2. To discover at what stage of the learning process changes in these systems occur.
3. To determine where in the brain (or nervous system) these neurochemical events occur.
4. To establish the behavioral specificity of the changes.

The methods employed by simple systems researchers are perhaps the most varied of the four strategies we have seen. In general, they focus less on behavior and emphasize the anatomical, electrophysiological, biochemical, and molecular techniques. The general approach is to study a behavior using the comparative method; to localize the site of the behavior using neuroanatomical techniques; and when the site(s) of behavioral change are located, to progress on to cellular and molecular analysis.

Behavioral–genetic analysis is an especially interesting tool for the simple systems researcher because the behavior of individual animals are conceived as behavioral "molecules." Techniques employed by the simple systems researcher include various behavioral procedures, neuroanatomical staining methods, recording of both electromyograms and intra- and extracellular potentials, patch and voltage clamping, cell fractionation, chromatography, electrophoresis, and spectrophotometry.

The contributions of the simple systems approach to the study of learning are many. It has stimulated the study of invertebrate behavior as a subject in its own right, increased our knowledge of invertebrate learning, and provided a cellular, and sometimes even a molecular, profile of many types of behavior. The success obtained with invertebrates has also stimulated the search for vertebrate models that are amenable to the same types of analysis.

Perhaps the greatest contribution of the simple systems approach is the possibility of creating a unified theory of behavior. Krasne (1984) commented on the relationship between invertebrate and vertebrate learning:

> Although vertebrates generally, and especially mammals, are more sophisticated learners than invertebrates, it would be unwise to assume that differences in abilities and in some of the phenomena of learning necessarily imply that different basic mechanisms are operating. The basic cellular processes responsible for coordination, integration, sensory information processing, production of movement, etc. seem to be very widespread amongst all animals with nervous systems. Therefore, in the absence of compelling evidence to the contrary we should also anticipate the generality of the basic cellular mechanisms that underlie learning abilities. (p. 72)

Hawkins and Kandel (1984) combined the results of the invertebrate and vertebrate data and have identified the following five features of learning shared by vertebrates and invertebrates at the cellular level:

1. Elementary aspects of learning are not diffusely distributed in the brain but can be localized to the activity of specific nerve cells.
2. Learning produces alterations in the membrane properties and in the synaptic connections of those cells.
3. The change in synaptic connections so far encountered have not involved formation of totally new synaptic contacts. Rather, they are achieved by modulating the amount of chemical transmitter released by presynaptic terminals of neurons.
4. In several instances, the molecular mechanisms of learning in-

volve intracellular second messengers and modulation of specific ion channels.

5. Molecular mechanisms of conditioning, an associative form of learning, are an elaboration of the same molecular mechanisms involved in sensitization, a nonassociative form of learning.

Whether the commonalities between invertebrates and vertebrates are more than superficial can be determined only by a judicious use of the comparative method and continued recognition—by government and private funding agencies—of the unique "partnership" between invertebrates and the many scientific disciplines that depend on them. Those interested in the simple systems approach will find much material in Table A-1 and chapter 7. Of special interest are a number of *Scientific American* articles that nicely describe the use of invertebrates such as *Drosophila* (Benzer, 1973), *Hermissenda* (Alkon, 1983; 1989), *Tritonia* (Willows, 1971), *Aplysia* (Kandel, 1970; Kennedy, 1967), and *Hirudo* (Nicholls & Van Essen, 1974) in the analysis of behavior.

How to Choose a Particular Research Strategy

For readers entering the study of invertebrate learning for the first time, the choice of which strategy to use in your own research can be daunting. The variety of strategies devised to uncover the mechanisms of invertebrate learning is one of the most conspicuous characteristics of the invertebrate learning literature. As the previous section illustrated, invertebrate learning has been the object of investigation from a wide range of scientific disciplines—from the molar approach to behavior analysis of the comparative psychologist to the more molecular strategy taken by those who follow the simple systems approach.

Before I discuss some guidelines used in selecting a particular strategy, it is important to impress upon you that the study of invertebrate learning is a cooperative enterprise. No one strategy is more important than any other. The four strategies we have discussed interact and influence each other with the common goal of understanding behavior. There are no better or worse strategies except as they are applied to particular research problems. Comparative psychologists, of which I am one, are among a group of behavioral scientists that have long provided the simple systems researcher with many of the behavioral techniques and much of the vocabulary necessary to detect and describe the learning ability of invertebrates. In turn, biologists, neuroethologists, and physiologists provide a molecular analysis of behavior that is of interest to comparative psychologists. Indeed, it is not uncommon to see psychologists working side by side with biochemists, entomologists, ethologists, geneticists, and physiologists (among others) as part of an interdisciplinary research team. The psychologist, behavioral biologist, and ethologist are critical to the success of the team because their presence ensures that the results of the "molecular gods" make contact with the behavior of a freely moving animal—not an isolated piece of tissue. This vital role played by behavioral scientists—especially comparative psychologists—can often be overlooked in the rush to uncover the underlying mechanisms of behavior.

The importance of behavioral studies has been eloquently expressed by Seymour Kety (1968), who stated

> It has most certainly not been my intention to deny the tremendous importance and major contributions which biochemistry and biophysics and the biological sciences generally have achieved within our lifetime. I have merely wanted to point out that we do not always get closer to the truth as we slice and homogenize and isolate, that what we gain in precision and in the rigorous control of variables we sometime lose in relevance to normal function, and that, in the case of certain diseases or problems, the fundamental process may often be lost in the cutting. A Heifitz and a Rubinstein playing different sonatas at the same time will produce a cacophony which the most exhaustive study of either individually would never have revealed, and a truer picture of the nervous system and behavior will emerge only from its study by a variety of disciplines and techniques, each with its own virtues and its own peculiar limitation. (p. 308)

If you have the opportunity to be part of such a research team, or observe one in action, take advantage of it. To do so will improve the quality of your research and give you the necessary experience that cannot be obtained from a book.

There are basically two ways to select a research strategy. First, the type of problem you select will often, at least in the initial stages of your research plan, dictate the type of strategy that will be used. If your problem, for instance, involves understanding the physiology of learning, you are obliged to learn all you can about the techniques, data, and philosophy associated with the simple systems approach. On the other hand, if your interest lies more toward the evolution of learning, an examination of the ethological and comparative psychological approaches is a good place to start. The disadvantage of this method of selecting a strategy is that you may not be competent to use a particular strategy without additional training. A psychologist and ethologist, for example, may be experts in the use of behavioral analysis and behavioral techniques but may not have the experience to use the tools of the neuroscientist. The converse is also true: A neuroscientist will not necessarily be aware of the nuances associated with behavioral analysis.

A second method of selecting a research strategy, which is perhaps more common than the first, is to approach the problem from your strength. Using this method, you will apply your own experiences and expertise to find a solution. For instance, if your scientific discipline is in the area of psychology, most likely you will use the tools and theoretical approach of the psychologist in tackling a research question. In my view the drawback here is that you limit yourself unnecessarily. Yes, there are certainly advantages to specialization, but there are also advantages to generalization—not the least of which is that your wide variety of knowledge will always be in demand.

However you select your strategy, always keep in mind that as a good researcher, you must consider how each strategy is applied to your research question. In other words, you must keep an open mind regarding the strengths and weaknesses of each strategy. Understanding the neuronal mechanisms of a particular behavior using the simple systems approach becomes much more exciting and relevant when you can understand how the mechanisms relate to the behavior of the whole animal (i.e., ethology and comparative psychology), how these mechanisms are similar and different in animals of the same and related species (i.e., comparative psychology), and how the behavior and its underlying mechanisms are passed on to future generations (i.e., behavioral genetics).

Some Necessary Skills for Invertebrate Research

In addition to having an open mind, a good invertebrate researcher will have a range of skills. More often than not, we are required to build apparatus, have a fair knowledge of electronics, and know our way around a computer. Often these skills are developed out of necessity because equipment is not commercially available or it must be modified in some way to do what we need it to do. Many of these skills can be learned by enrolling in the relevant course (i.e., use your elective), through volunteer efforts, or on your own. An added advantage of having a range of skills is that as scientists, we can control at least some of our equipment costs, and we are no longer limited to what someone else thought of beforehand.

Following is a list of skills that I have found useful in the study of invertebrate learning and encourage you to obtain:

1. **Learn to communicate with your colleagues**. As a scientist, you will be required to communicate your findings both orally and through writing. If you have a problem with public speaking, get over it. (One method I have found useful is to practice in front of a mirror.) Keep in mind that as a beginning student, you will be expected to make mistakes. However, as you advance in your career, the expectation changes. In addition to communicating your findings to colleagues, communication skills are necessary when contacting prospective colleagues. Do not be afraid to contact those scientists in whose work you are interested.

2. **Acquire a background in zoology and physiology**. Make an effort to "know your animal." Attempt to understand how your invertebrate subject is similar to and different from others. An understanding of the natural history and physiology of your animal is of critical importance in the design and interpretation of your experiment.

3. **Acquire a background in electronics**. Learn how to read a circuit diagram and construct basic circuits from those diagrams. It is also useful to know how to use a volt meter, oscilloscope, soldering iron, and other standard electronic equipment.

4. **Acquire a background in basic shop skills**. Learn how to use such power tools as a drill, table saw, lathe, and milling machine. Such skills come in handy when you are called upon to build an assortment of invertebrate-sized mazes and other miniature devices.

5. **Acquire a background in computers and instrumentation**. Learn how to use a computer, and familarize yourself with the various software packages. The more software packages you become familar with, the easier it becomes to learn a new one. There are a number of sophisticated software and hardware packages available for the analysis and design of experiments. Acquiring the most sophisticated equipment available is nice, but your work station will only collect dust if you do not know how to use it.

6. **Acquire good library research skills**. Make the library one of your best friends. Introduce yourself to the library staff, and familiarize yourself with the library's contents. Learn where the invertebrate books and relevant journals are located. It is espe-

cially important to learn how to use the electronic databases such as PsycLIT and *Index Medicus*.

7. **Learn organization skills**. Good organization skills are a must. Proper organization means never having to waste valuable time searching for information. Learn how to keep a laboratory notebook and record your observations. Do not make the mistake of thinking that you will remember the relevant details of a particular project; in the span of 2 weeks you will have forgotten what you did.

8. **Learn basic photography and video techniques**. Obtaining a working knowledge of photography and video making is extremely useful in creating graphic presentations for conferences and research reports and recording the results of your experiments on tape.

Other issues you must consider when selecting a research strategy are somewhat intangible. Be enthusiastic and enjoy what you are doing. Have confidence in what you are doing and in your own ability. The process of discovery is never an easy one, especially when working with invertebrates. If you are easily discouraged or unenthusiastic about employing a particular strategy, do not use it. To do so would risk the possibility that you would not make your best effort and become sloppy in the execution of your experiment. Moreover, when you are considering employing a strategy that is other than what you are accustomed to, resist the tendency to criticize. Keep an open mind. As a personal example, for many years, I did not have a high regard for ethologists. However, when I started meeting some "up close and personal," my attitude toward ethological analysis began to change to the point that I now employ some ethological concepts in my own research.

Practical Considerations in Choosing a Research Strategy

In addition to the appropriate scientific background and range of skills and the choice of a problem, there are some practical issues that must be considered before selecting a particular strategy. Following are some issues that I have found useful:

1. **How much money will it cost to pursue your research question?** If you are interested in exploring the neuronal mechanisms of learning in snails, for instance, a physiological work station can easily cost between $10,000–$15,000. Purchasing and maintain-

ing snails will run a few hundred dollars, and the apparatus needed to measure learning may cost several hundred dollars.

2. **Do you have the necessary apparatus and animal facilities?** Be advised to take time to learn how to keep your animals healthy in the laboratory. Appendix A lists reference materials that describe how to house a variety of invertebrates. An added complication is that, unlike the apparatus available for the study of vertebrates, there are few commercially available apparatus for the study of invertebrate learning. Thus, you will need access to construction materials and tools, and you must be prepared to construct your own mazes, choice chambers, and other necessary equipment. In the next chapter, I will discuss the factors that must be considered in designing an invertebrate learning apparatus.

3. **Do you have the necessary experience working within a particular strategy?** Resist the temptation to jump into a new area of research without proper training or supervision. Seek out an advisor who can teach you the necessary skills.

4. **Do you have the necessary experience working with a particular invertebrate?** If not, seek advice.

5. **What are your strengths and weaknesses?** Give yourself an honest evaluation. Everyone has strengths and weaknesses.

6. **Do you have access to a research team?** Belonging to a research team is extremely important. The group environment encourages a fresh outlook that often aids in the solution of a problem. The advantages of working in a cooperative atmosphere should not be overlooked. In addition to the emotional support a research team offers, there are practical benefits as well. One of the most obvious is that you are surrounded by people who possess skills that you may not have. Given today's explosion of knowledge, it is unlikely that one person can acquire in a reasonable time all the detailed information and skills necessary to conduct different kinds of invertebrate learning experiments. Members of an invertebrate research team often include a biochemist, zoologist, psychologist, and physiologist. In addition, the team often has access to a mechanical shop, electronics shop, computer programmer, and an electron microscope. In such an arrangement, you are able to tackle a research question from any combination of the strategies we discussed earlier.

Often, research teams are established with the assistance of grant money. Grant money is given by the federal government, private foundations, and businesses to assist you in finding answers to specific questions. This money enables you to hire individuals and purchase equipment that can assist you in answering your particular research question. Money is not the only way to establish a research team, however. As faculty interested in entering the area of invertebrate learning, members of the team can be recruited through various university and college departments. It has been my experience that a good idea or interesting problem will always attract its share of eager participants from, for example, the departments of biology and entomology as well as from nearby educational institutions. This same strategy can be used if you are a student. For example, as an undergraduate student, I was able to convince an engineering student to devote his senior project to an area that would benefit his project and mine; you can do the same. The idea—whether you are a faculty member on a tight budget or a student—is to surround yourself with a circle of friends and colleagues who share your interest in a particular question.

The Varieties of Learning: An Overview

A common theme running throughout the four strategies we have discussed is the topic of problems of learning. In this section, I will provide a brief overview of the different types of learning in order to set the stage for the detailed discussions in subsequent chapters. A more detailed overview will be available at the beginning of chapters 4, 5, and 6. These different procedures will be familiar to those readers who have taken an introductory psychology course or a course on learning.

Before you can conduct an experiment on invertebrate learning, it is necessary to familiarize yourself with the some of the basic terminology and paradigms. Like any scientific discipline, the study of learning contains its own specialized terminology. Mastering such terminology is essential if you are to communicate effectively with colleagues and correctly interpret the results from research reports. If you are a psychologist, you will find much of this vocabulary familiar. This is not by chance. As I have mentioned throughout this chapter, the study of learning is shared by many disciplines, including biochemistry, neuroscience, and entomology. Nevertheless, much of the vocabulary of invertebrate learning is taken from the vocabulary of the psychologist. This is due primarily to the influence of such pioneers as Ebbinghaus, Thorndike, Pavlov, Watson, Schneirla, Hull, Tolman, and Skinner. Moreover, many of those who followed and made significant contributions to the study of invertebrate learning, such as M. E. Bitterman, Jerry Hirsch, Stanley Ratner, James McConnell, E. R. John, Philip Applewhite, W. C. Corning, and Frank Krasne, were trained as psychologists. To better understand the process of learning and uncover the underlying mechanisms, psychologists have divided the categories of learning into nonassociative and associative. I will discuss each one in turn.

Nonassociative Learning

Nonassociative learning is a form of behavior modification involving the association of one event, as when the repeated presentation of a stimulus leads to an alteration of the probability or strength of a response. Nonassociative learning is considered to be the most basic of the learning processes. The animal does not learn to do anything new or better; rather, the innate response to a situation or a particular stimulus is modified. The two types of nonassociative learning that have received the most analysis are habituation and sensitization.

Habituation

Habituation refers to the reduction in responding to a stimulus as it is repeated. For a decline in responsiveness to be considered an instance of nonassociative learning, it must be determined that sensory adaptation and motor fatigue do not exert an influence. There are generally two types of habituation recognized: short-term and long-term. The principal difference is the length of memory. Studies of habituation show that it has several characteristics, including the following:

1. The more rapid the rate of stimulation is, the faster the habituation is.
2. The weaker the stimulus is, the faster the habituation is.
3. Habituation to one stimulus will produce habituation to similar stimuli.
4. Withholding the stimulus for a long period of time will lead to the recovery of the response.

Sensitization

Sensitization refers to the augmentation of a response to a stimulus. In essence, it is the opposite of habituation and refers to an increase in the frequency or probability of a response. There are also two types of sensitization: long-term and short-term. Studies of sensitization show that it has several characteristics, including the following:

1. The stronger the stimulus is, the greater the probability is that sensitization will be produced.
2. Sensitization to one stimulus will produce sensitization to similar stimuli.
3. Repeated presentations of the sensitizing stimulus tend to diminish its effect.

Associative Learning

Associative learning is a form of behavior modification involving the association of two or more events, such as between two stimuli, or between a stimulus and a response. In associative learning, an animal does learn to do something new or better. Associative learning differs from nonassociative learning by the number and kind of events that are learned and how the events are learned. Another difference between the two forms of learning is that nonassociative learning is considered to be a more fundamental mechanism for behavior modification than those mechanisms involved in associative learning. This can easily be seen as we move through the animal kingdom. Habituation and sensitization are present in all animal groups, but classical and operant conditioning are not (Razran, 1971). In addition, the available evidence suggests that the behavioral and cellular mechanisms uncovered for nonassociative learning may serve as the building blocks for the type of complex behavior characteristic of associative learning (Groves & Thompson, 1970; Hawkins & Kandel, 1984; Razran, 1971). The term *associative learning* is reserved for a wide variety of classical, instrumental, and operant procedures in which responses are associated with stimuli, consequences, and other responses.

Classical Conditioning

Classical conditioning refers to the modification of behavior in which an originally neutral stimulus—known as a *conditioned stimulus* (CS)—is paired with a second stimulus that elicits a particular response—known as the *unconditioned stimulus* (US). The response which the US elicits is known as the *unconditioned response* (UR). An organism exposed to repeated pairings of the CS and the US will often respond to the originally neutral stimulus as it did to the US. There are two major classes of classical conditioning experiment. If, for example, a honeybee is trained to extend its proboscis to an odor that is followed by sugar water, we have appetitive classical conditioning. If the odor is followed by a brief aversive event such as shock, we have aversive, or as it is sometimes called, defensive conditioning. Studies of classical conditioning show that it has several characteristics, including the following:

1. In general, the more intense the CS is, the greater the effectiveness of training is.
2. In general, the more intense the US is, the greater the effectiveness of training is.
3. In general, the shorter the interval is between the CS and the US, the greater the effectiveness of training is.
4. In general, the more pairings there are of the CS and the US, the greater the effectiveness of training is.
5. When the US no longer follows the CS, the conditioned response gradually becomes weaker over time and eventually stops occurring.
6. When a conditioned response has been established to a particular CS, stimuli similar to the CS may elicit the response.

Instrumental and Operant Conditioning

Instrumental and *operant conditioning* refer to the modification of behavior involving an organism's responses and the consequences of those responses. It may be helpful for you to conceptualize an operant and instrumental conditioning experiment as a classical conditioning experiment in which the sequence of stimuli and reward is controlled by the behavior of the animal. In contemporary usage, the terms *instrumental* and *operant conditioning* are used interchangeably. However, there are several differences in terms of methods and procedures. These differences as well as data generated by these procedures will be discussed in chapter 6. Instrumental and operant conditioning fall into four major categories. These categories are familiar to those readers who have taken an introductory psychology course or a course on learning.

When there is a positive relationship between a response and a desirable outcome, such as when a cockroach finds a bit of pineapple at the end of a maze, the phenomenon is known as *reward training*. A special case of reward training is known as *escape*. In escape training, a response terminates an unwanted event. For example, the cockroach may run through the maze to escape a bright light. The reward in this situation is finding a dark compartment at the end of the maze. If the roach runs through a dark maze only to find a lighted compartment and then over time it refuses to run the maze, this is an example of what is known as *punishment training*, or *passive avoidance*. When the door of the maze is raised and the roach is required to run the maze to prevent the onset of light, what is demonstrated is known as *signaled avoidance conditioning*. The signal in this example is a compound of the noise and vibration associated with the opening of the door. Studies of instrumental and operant conditioning show that they have several characterisitics, including the following:

1. In general, the greater the amount and quality of the reward are, the faster the acquisition is.
2. In general, the greater the interval of time is between response and reward, the slower the acquisition is.
3. In general, the greater the deprivation level is, the more vigorous the response is.
4. In general, when reward no longer follows the response, the response gradually becomes weaker over time and eventually stops occurring.

A Note on Apparatus

The apparatus used to measure and study nonassociative and associative learning is not complex at the behavioral level. For example, an apparatus to study habituation in a hermit crab can be as simple as waving your hand over the crab and observing whether it retreats into its shell. Instrumental studies of honeybee learning routinely use an assortment of colored plastics and petri dishes. Conditioning of the proboscis, or "trunk," of an insect can easily be carried out by confining an insect such as a fly to a small tube and administering an odor signal using a plastic syringe. Some of the more familiar apparatus include miniature versions of mazes and running wheels. Much more sophistication is required when you measure bioelectrical signals associated with the activity of nerves. To measure such activity, you will need an assortment of fine wires or fluid-filled glass tubes known as *electrodes* and some way to amplify the small electrical signals so that they can be seen and recorded. This is done with the help of a device known as a *preamplifier*. A discussion of apparatus used in electrophysiology is beyond the scope of this book. However, because the design of behavioral apparatus is so critical to the success and interpretation of invertebrate learning experiments, I will devote the next chapter to it.

The Comparative Method

In addition to the use of learning paradigms, a second theme linking the various invertebrate strategies is the *comparative method*, which is fundamental to each of the strategies we have examined. In this section, we will take a look at the comparative method and see how this method can be applied to your own research.

The comparative method is well known to those readers schooled in biology, ethology, physiology, and zoology; indeed, it is characteristic of all scientific disciplines, including psychology. Stanley Ratner (Denny & Ratner, 1970; Ratner, 1980) has developed a convenient scheme that divides the comparative method into six sequential stages. Each stage depends on the previous one, and developments in one stage influence not only the subsequent phase but the preceding phase as well. Although his scheme was not developed specifically for invertebrate learning, you will find it useful as a guide in conducting your research on invertebrates.

Stage 1: Background Information

The goal of invertebrate learning research is, obviously, to understand how these animals learn. Before this goal can be reached, you must have information about the natural history of the animal as well as physiological and zoological data—in other words, background information.

Materials on the behavior, natural history, and physiology of many invertebrates are easily found in any textbook on invertebrate zoology such as that by Lutz (1986) or by Pearse, Pearse, Buchsbaum, and Buchsbaum (1987). Two excellent books describing invertebrate nervous systems and their relationships to vertebrate nervous systems are by Bullock et al. (1977), and by Shepherd (1988). If you are seriously interested in conducting invertebrate research, purchase one or more textbooks describing invertebrate zoology and physiology.

The most comprehensive review of the invertebrate work conducted prior to 1940 appears in Warden et al.'s (1940) *Comparative Psychology: Plants and Invertebrates*. This volume, which is the second book of their three-volume series on comparative psychology, contains a marvelous review of behavior modification in plants such as the sensitive plant *Mi-*

mosa and in invertebrates ranging from single-cell animals to worms to social insects. Diagrams are also presented that illustrate the invertebrate apparatus of the period, and there are photographs of the early pioneers of invertebrate learning. The definitive review of research conducted between 1940 and 1972 appears in Corning, Dyal, and Willows' three-volume set (1973/1975). Since 1972, there has been no comprehensive review of studies of invertebrate learning. Rather, a number of specialized review articles have appeared. The list includes reviews by Krasne (1984), Farley and Alkon (1985), Carew and Sahley (1986), Byrne (1987), Bitterman (1988), and Smith and Abramson (1992).

Information can also be obtained by accessing any of the computer database services such as *Index Medicus*. Take advantage of these services. They are simple to use and can be a powerful research tool. You should be aware, however, that a computer search has its limitations. Articles appearing before 1966, for instance, are not part of the *Index Medicus* database. In addition, the services can be somewhat ineffective (and costly) if you wish to "browse" through the literature. In many libraries these services are now free. Furthermore, the database you select may not contain relevant books and journals; determine the scope of the database before you begin and make sure it contains a scan of biological journals. As a test for accuracy, have several citations handy that the computer search should detect. If these articles are not on your printout, consider using another database.

Your own experiences and powers of observation are another source of information. When working with a new animal, always take time to observe it, particularly under natural conditions. If this is not practical, obtain data from ethological reports and zoological texts. Also, consider contacting any scientist whose research interests you. Often, you can obtain information and tips not available in a publication. Write to these scientists. You will probably find that we are a friendly lot. Speaking for myself, I would certainly enjoy hearing from you. Appendix B contains the names and addresses of scientists involved with invertebrate learning. Review articles describing the learning ability of some of the more popular invertebrates can be found in Table A-1.

Stage 2: Classification of Behavior

When you have a sufficient amount of information on your invertebrate subject, you can begin the difficult task of classifying this information into meaningful categories. Classical, instrumental, and operant conditioning are well-known categories of associative learning (see pp. 38–40).

The problem of classification is one of the more difficult you will face in the study of invertebrate learning, and it will be discussed in chapter 8. The major difficulty is the lack of an accepted classification of behavior. Those of us who work in the area of invertebrate learning do not always agree, for example, on such basic issues as what constitutes classical conditioning, whether there is a difference between operant and instrumental conditioning, and whether it is possible to compare and rank animals on the basis of their ability to learn. Most of us in the field, however, believe this controversy to be a temporary situation. As any introductory textbook in biology or chemistry will show, there was a time when naturalists and comparative anatomists worked without an evolutionary classification and chemists worked without a periodic table of elements. Remember that the study of learning is only slightly over 100 years old, and despite the complex issues involved, one hopes that a classification of behavior will be in place during your lifetime if not mine. For a discussion of the issues involved in the classification of behavior, see Dyal and Corning (1973), Hodos and Campbell (1990), and the discussion in chapter 8.

Stage 3: Research Preparations

Once you have the necessary background information and have placed this information into categories, you can begin to design research preparations. These model the behavior identified in Stage 2. The term *preparation* is frequently used by experimental biologists and zoologists as well as those involved in the simple systems strategy to refer to the biological object under study. For example, we speak of a "crab preparation" or a "honeybee preparation." For our purposes, the term *preparation* will refer to the subject and procedure of your experiment. Thus, if you are studying classical conditioning in the crab, you have a "crab classical conditioning preparation." The discovery of a preparation is rather exciting. It means that following many failures, you have found a stable experimental situation in which to manipulate a host of variables.

The models created by such preparations are one of the outstanding contributions of the invertebrate learning literature. This is a particular characteristic of the simple systems approach to behavior analysis. If you are interested in, for example, studying classical conditioning of insects, the olfactory conditioning in the honeybee is an excellent model system.

Hermissenda and *Aplysia* are good model systems for those interested in the study of learning in molluscs. The discoverers of model systems hope that their results can be generalized to other, and perhaps more complex models, both invertebrate and vertebrate. Often this hope is justified only after years of intensive investigation.

As Ratner (1980) has discussed, a good model system possesses four characteristics. First, the model must be a valid representation of the behavior you wish to study. Determining this validity is not easy without an accepted behavioral classification. Second, your preparation must yield reliable and reproducible data. The invertebrate learning literature is full of experiments that have not been replicated. Third, the preparation must represent an unambiguous example of the behavior in question. If you are interested in modeling classical conditioning, ensure that the model is not contaminated by the presence of nonassociative or operant effects. The fourth characteristic is one of convenience. The preparation should be easy to use. Table A-2 provides a listing of articles that provide detailed information on many of the available models.

To help you evaluate and classify invertebrate experiments, I have slightly modified the guidelines used by Townsend (1953) in his "Form for Planning or Reporting Experimentation" (adapted in Exhibit 2-1). These may be photocopied by students and teachers for classroom use. The guidelines will direct you to the major points that must be considered when designing your own invertebrate experiment and evaluating those you find in the literature.

Stage 4: Parametric Analysis

Having developed a research preparation that is both valid and reliable, that offers a clear example of the behavior under study, and that is convenient to use, you can begin the sometimes arduous task of parametric analysis. Although the identification and manipulation of variables that affect the invertebrate preparations is not glamorous, it is necessary. In this stage, you systematically and carefully explore the factors that influence our conditioning situation, or conditioning preparation. Often, such analysis takes the form of manipulating variables known to influence the specific type of learning or physiological mechanism of interest.

The task of identifying variables that may affect your preparation is complicated somewhat by the likelihood that several varaibles may interact. One of the principal benefits gained from the work of comparative psychologists, entomologists, ethologists, and naturalists is that they have

Exhibit 2-1

Guidelines for Planning or Reporting Experimentation

1. What is the problem?
2. State the problem in terms of a hypothesis.
3. Provide background information on the problem.
4. What is the independent variable?
5. What is the dependent variable?
6. How is (are) the dependent variable(s) to be measured?
7. Relate the procedure to a behavioral category.
8. What are the similarities and differences between this preparation and analogous invertebrate and vertebrate preparations?
9. What controls are necessary? How are they implemented? Why do you need them?
10. What is the procedure to be followed in conducting the experiment?
    a. Provide a description of the natural history and physiology of the subject.
    b. Diagram the apparatus.
    c. Describe exactly what you plan to do.
    d. Describe how you plan to analyze the research results.
11. Review the research design.
    a. What results, if obtained, would support the hypothesis?
    b. What results, if obtained, would fail to support the hypothesis?
    c. Does the preparation accurately reflect the response class supposedly under investigation?
    d. Does the design lead to a clear example of the behavior under investigation?
12. Conduct the experiment.
    a. Describe unplanned occurrences that were present and that may have influenced the results.
    b. Describe the behavior of individual subjects.
    c. Summarize the research results in tables, graphs, and/or other clear means of presentation.
13. Interpret the results.
    a. Describe the tables and graphs and statistical analysis from the point of view of proving or disproving the hypothesis.
    b. Determine whether the experiment yielded reliable information about the response system under investigation.
    c. Present samples of individual data.
    d. State the conclusions.

identified many of these variables. Some of the important organismic variables include age, sex, developmental stage, motivational level, and biological rhythms. Variables known to influence learning procedures include amount, probability, frequency, and delay of reinforcement; and type of response and the effort required to emit or elicit it. Other variables of importance include the duration, intensity, and temporal order of stimuli used during training. As you can see from this short list, the number of variables that can potentially interact with your preparation seem almost endless. Nevertheless, these variables must be assessed, and a thorough parametric analysis is the only way to do so.

Stage 5: Comparison and Relations

One of the more exciting aspects of the comparative method is the ability to compare your preparation with others and to discover relations between them. For example, you may wish to compare the leg lift learning of an intact cockroach with a headless roach or with a fly, a locust, or even a crab. This is most effectively done after you have established a behavioral taxonomy (Stage 2) and created a powerful research preparation (Stages 3 and 4). Comparisons can be made at the level of the nervous system as in comparing the nerve net of a hydra with the more complex systems found in advanced flatworms and arthropods. The comparison can also be made at the level of behavior as when we compare the learning of animals across the evolutionary scale.

Stage 6: General Theory

Perhaps your most difficult task is found in Stage 6, where you forge your invertebrate data into a general theory that can account for the learned behavior of a variety of organisms. In the final stage, theoretical mechanisms are created that relate the various classes of behavior to each other and, where possible, to the underlying cellular mechanism(s). Such theories attempt the noble task of integrating a wide range of often complicated behavioral and physiological data into a general picture of how we learn. Such grand theories have been attempted in the past by many of the giants in learning research such as Pavlov, Skinner, Hull, and Kornorski. The attempts met with limited success, and unfortunately, additional attempts were abandoned for many years.

Over the last two decades, invertebrate learning research and the development of neural network models have stimulated a renewed search for general mechanisms. (For a review of neural networks, see Levine [1983, 1989], Tesauro [1986], Klopf [1988], and Hawkins & Bower [1989].) Hawkins and Kandel (1984) suggest, for example, that learning in vertebrates and invertebrates shares such common features as the involvement of second messengers and modulation of specific ion channels. If additional similarities can be identified, you will be that much closer to what Clark Hull (1945) envisioned almost 50 years ago: that species differences may be reflected by the parameters in equations that describe behavior.

Summary

In this chapter, I have discussed four major strategies used in the analysis of learning. We have seen how these strategies interact and influence one another. In addition, I have discussed guidelines that are useful in selecting a particular strategy and provided an overview of the major categories of nonassociative and associative learning. To help you in your own research, a list of recommended skills was provided as well as guidelines that can assist you in designing your own experiment and analyzing others already available.

Discussion Questions

- Compare and contrast four strategies associated with invertebrate learning.
- Of the four strategies, discuss your personal favorite and your least favorite.
- What is the difference between nonassociative and associative learning?
- Why is it important to construct behavioral categories?
- Provide an overview of the major features of classical, instrumental, and operant conditioning.
- Describe the major features of the comparative method.
- Of the recommended skills found to be important for an invertebrate researcher, how many do you possess?
- Using the Guidelines for Planning or Reporting Experimentation,

visit the library, find a paper on invertebrate learning, and analyze according to the guidelines.

- Using the Guidelines for Planning or Reporting Experimentation, conduct a "thought experiment" on classical conditioning, and analyze according to the guidelines.
- Describe and provide an example of an ethogram.

3 Tools of the Trade

Preview Questions

- What are the characteristics of a good invertebrate learning apparatus?
- Where can I find valuable construction tips?
- What are the most widely used invertebrate learning apparatus?
- Describe the following apparatus and give some tips for their use:
    1. activity wheel
    2. actograph
    3. maze
    4. runway
    5. Skinner box
    6. shuttle box

- What apparatus are best used for the following procedures, and what are some tips for their use?
    1. the study of food aversion and food preferences
    2. the study of proboscis conditioning
    3. the Horridge procedure
    4. the free-flying procedure
    5. the genetic dissection of learning
    6. the classical conditioning of contraction

Before undertaking a study of invertebrate learning, it is necessary to familiarize yourself with not only the data and procedures but also the apparatus with which the experiments are actually conducted. It is my

Table 3-1

Apparatus and Their Functions

Apparatus	An.	Phys.[a]	Habit.[b]	Sens.[b]	Class.[b]	Instr.[b]	Oper.[b]
Activity wheel	V[d]	X	X				
Actograph	V	X	X	X	X		
Maze	V					X	
Runway	V					X	
Lever press	V					X	X
Shuttle box	V					X	
Food prefer- ence	V	X			X		
Olfactory condition- ing	V	X			X	X	
Horridge procedure	V	X				X	
Free-flying technique	V				X	X	
Counter- current device		X			X	X	
Withdrawal condition- ing		X	X		X		

continues

Note. [a]Device or technique has been used for the physiological analysis of invertebrate learning. [b]Category of learning in which device has been used in invertebrate research. [c]Specific procedure in which device has been used in invertebrate research. [d]Device or technique is used also for vertebrates. An. = Animal; Phys. = Physiological; Habit. =

view that the new generation of students interested in invertebrate learning has lost the art of designing behavioral apparatus. The goal of this chapter is to contribute to reversing this trend. The chapter is constructed in such a way that you will see many examples of invertebrate apparatus and how they are applied to various learning procedures and obtain some tips on how they are used. In many cases an apparatus can be used for more than one type of conditioning situation. For example, an apparatus design for the study of classical conditioning can be adapted for the study of instrumental conditioning. To help you select an apparatus for a particular conditioning procedure, Table 3-1 indicates how the apparatus

Table 3-1, continued

Apparatus	Avoid.[c]	Choice[c]	Discr.[c]	Esc.[c]	Gen.[c]	Pun.[c]	Sched.[c]
Activity wheel			X	X			
Actograph	X		X	X			
Maze	X	X	X	X	-	X	
Runway			X	X		X	
Lever press			X			X	X
Shuttle box	X	X	X	X	X	X	
Food pref- erence	X	X	X		X	X	
Olfactory condition- ing	X	X	X	X	X	X	
Horridge proce- dure	X		X	X		X	
Free-flying technique	X	X	X	X	X	X	
Counter- current device	X	X	X		X	X	
Withdrawal condition- ing	X		X		X		

Habituation; Sens. = Sensitization; Class. = Classical; Instr. = Instrumental; Oper. = Operant; Avoid. = Avoidance; Discr. = Discrimination; Esc. = Escape; Gen. = Generalization; Pun. = Punishment; Sched. = Schedules.

has been used in the past. I hope that you will agree that the apparatus developed to measure the learning of invertebrates are varied and often ingenious. They are certainly the result of much effort and observation.

Behavioral apparatus range from simple runways designed to measure, for instance, the speed of a cockroach as it hurtles toward a piece of pineapple to sophisticated electronic devices sensitive enough to measure the bite of a tick. In this chapter, I will identify some of the more popular behavioral apparatus and techniques developed to detect the learning ability of invertebrates. Table A-3 provides a list of review articles and apparatus that are available for several species.

A well-designed apparatus takes advantage of the animal's natural history, selecting the most efficacious combinations of physical dimensions, material, stimuli, and independent variables. Such an apparatus does not influence the behavior under study, permits you to present stimuli and record behavior reliably, and is easy to use. For many researchers accustomed to the convenience of vertebrate apparatus, invertebrate apparatus may present some difficulties, the principal problems being a lack of standardization and the scarcity of commercially available designs. Be prepared to construct your own or to pay to have it built. If you pay for a device, be prepared for quite an expense. Be aware also that there is a lack of information regarding the reliability of invertebrate apparatus and the relationship between the techniques. Kirk and Thompson (1967), for example, provide the only data describing the factors that contribute to the performance of earthworms in a runway; Ratner (1968) provides some information on the reliability of the indices used in earthworm and planarian learning; Hay and Crossley (1977) discuss the importance of maze design in the analysis of *Drosophila* behavior; and Ricker, Hirsch, Holliday, and Vargo (1986) discuss a variety of factors that must be taken into account when designing proboscis conditioning experiments. Vertebrate apparatus has become so familiar that it is hard to imagine a time when determining reliability scores for various apparatus was a significant area of research, but this was the case until the 1950s. See Hilgard (1951) for a review of techniques used for determining reliability scores.

Construction Tips

Before investing time and money in an apparatus, consult those familiar with its use (see appendix B for some names and addresses). You will gain valuable tips not found in the published reports. Some construction tips can of course be found by consulting the references that describe the technique of interest. My invertebrate learning manual (Abramson, 1990) provides instructions for building low-cost versions of many of the most popular devices. Miller (1979) is an especially good source for tips on building actographs and designing free-moving and tethered insect preparations and isolated organ and tissue preparations. This book is well worth acquiring. Martin and Bateson (1986) is a good introduction to behavioral measurement and will make a fine addition to your personal library.

To learn how to automate your device, I recommend acquiring a book describing optoelectronic circuits. Radio Shack publishes a little book, *Engineer's Mini-Notebook*, that is full of inexpensive and useful photocell circuits. Many of these circuits can be used in actographs, shuttle boxes, and runways and to record small movements of individual appendages. It is important also to acquire a book on the use of strain gauges and transducers. Such a book is available from the firm Omega and is entitled *The Pressure Strain and Force Handbook*. It is free for the asking; the customer service number is (800) 622-2378. The *Journal of Experimental Biology*, *Behavior Research Methods, Instruments, and Computers*, and *Physiology and Behavior* are a few of the journals that regularly publish apparatus, circuits, and computer programs. It is also worth going through a copy of *Popular Electronics*. A particularly useful circuit for recording small changes in the position of an appendage is available from Miall and Hereward (1988). Their position transducer can be used, for example, to record the eye movements of a crab, or the leg movements of a locust.

When adapting a vertebrate apparatus for invertebrates, keep in mind that you are not simply making a miniature as you would furniture for a dollhouse. Incorporate into your devices the ability to easily remove pheromones, contaminants, and if need be, the animal from the apparatus. There is nothing more frustrating to the experimenter and the subject than when you have to bang the device against a table to extricate the animal. In addition, invertebrate learning and behavior is extremely sensitive to temperature and humidity changes, and therefore it may be necessary to build into your device some type of temperature and humidity regulator. You should also be aware that the behavior captured by your device may be difficult to interpret. For example, if you decide to study instrumental behavior in a maze or runway, be aware that the behavior may be produced by classical conditioning of approach responses, the interaction of classical and instrumental conditioning, or both. This problem is common to both vertebrate and invertebrate devices and can be solved by proper experimental design.

Invertebrates will present challenging design problems. There is perhaps no better way to "know your animal" than to take the time to construct apparatus to detect its behavior. How will you, for example, keep an earthworm moist when it is inside of a running wheel, keep a restrained bee from satiating during a classical conditioning experiment, place an ant gently inside of a shuttle box, or keep your apparatus free from disturbances such as vibration? How will you keep your subject from crawling on the walls and ceiling of the apparatus? How will you administer electric shock to an insect? If you shock an aquatic invertebrate, how do you ensure that the current does not alter the aquatic environment? From what material should the apparatus (and home environment) be constructed? Consider also the problem of identifying individual subjects. Shall you glue individual tags to them, and if so, what type of adhesive do you use? Will you restrict your animals to individual cages? If so, what is the effect of isolation on behavior? How will you determine the health of your subjects? For vertebrates, a clue can often be found by investigating their fecal matter. Have you ever tried to find the fecal pellet of an ant, let alone analyze it? When the experiment is complete, what will become of your faithful subjects? Depending on the sexual behavior of your subjects, you can have quite a problem on your hands. A fertilized female cricket, for example, can lay as many as 2600 eggs during its life span.

In the course of your original investigations of invertebrate learning, you may decide that the experimental design calls for new techniques. Resist the tendency to slavishly copy vertebrate apparatus. The vertebrate designs should be thought of only as a convenient starting point, for in the course of your investigation, the opportunity may arise to develop something unique. Such was the case with a signaled avoidance situation I helped develop for the free-flying honeybee (Abramson, 1986). Having been frustrated in my attempts to study signaled avoidance behavior by confining bees to shuttle boxes, I achieved success by modifying von Frisch's free-flying technique. This remains the only signaled avoidance situation in which the subject is free to leave the experiment at any time. A better example is the countercurrent apparatus developed by Benzer (1967) to measure behavior in groups of fruit flies. This apparatus— which has no known counterpart in the vertebrate literature—has now become a standard technique in the genetic dissection of behavior.

If your design calls for using existing invertebrate apparatus, keep in mind that it may need modernization. Many of the invertebrate techniques are more than 20 years old and are unautomated. This is particularly true of the earthworm and planarian classical conditioning techniques. It is also possible to incorporate physiological techniques into your device. If your design, for instance, calls for a cockroach shuttle box, it is a relatively simple matter to record electromyograms from the roach leg. Oakley and Schafer (1978) is an excellent introduction to these techniques.

Apparatus for the Study of Invertebrate Learning

The remainder of this chapter describes some of the most widely used invertebrate devices and techniques. The apparatus most closely resembling vertebrate designs are presented first, followed by those devices and techniques uniquely suited for invertebrates. No attempt was made to categorize the devices according to function. You will often find that a device designed for one purpose can be adapted for many other purposes. The classical conditioning situation developed for earthworms and planarians, for example, can easily be modified to study habituation, sensitization, discrimination, and other behavioral phenomena. Each apparatus is described in the list that follows, and typically, several examples are given. Table 3-1 allows you to find out quickly how the apparatus is used.

Activity Wheel

An *activity wheel* (also known as a *running wheel* or *revolving drum*) is a cage built in the shape of a wheel to measure the activity of an animal. The animal is confined inside and by moving, turns the wheel. Some models tether the animal to a rod that permits the animal to move on the rim rather than inside the wheel. Measures of activity are easily obtained by counting the number of rotations or distance traveled. A running total or "cumulative record" can also be easily calculated. The apparatus is useful when gross patterns of behavior are of interest, particularly when the animal is tested over several hours or days. A drawback of the device is that it is not useful if precise locomotor sequences are desired and the animal must be isolated, possibly distorting the results. Moreover, the vertebrate literature suggests that the amount of activity may be over-estimated because the movement of the wheel itself will encourage further running. Whether this factor operates on invertebrates is not known. The running wheel is often used to assess the influence of drugs and to detect rhythmic behavior. The device has seen limited application in studies of learning and memory.

Running wheels are available for many invertebrates, including earthworms (Marian & Abramson, 1982), fruit flies (DeJianne, McGuire, & Pruzan-Hotchkiss, 1985), houseflies (Miller, Bruner, & Fukuto, 1971), roaches (Ball, 1972), and slugs (Sokolove, Beiswanger, Prior, & Gelperin, 1977). Figure 3-1 depicts a conventional running wheel for the earthworm (Marian & Abramson, 1982). The animal is confined to a V-shaped depression within the wheel. Holes located on the outside of the wheel allow air to circulate through the depression. Figure 3-2 depicts a wheel for the housefly (Miller et al., 1971). In this device, the animal is not

Figure 3-1

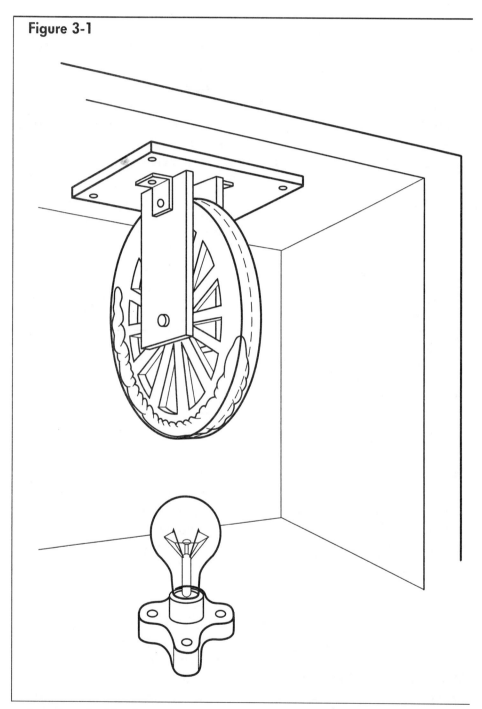

A running wheel for the earthworm. From "Earthworm Behavior in a Modified Running Wheel" by R. W. Marian and C. I. Abramson, 1982, *Journal of Mind and Behavior*, *3*, p. 68. Copyright 1982 by the Institute of Mind and Behavior. Adapted with permission.

Figure 3-2

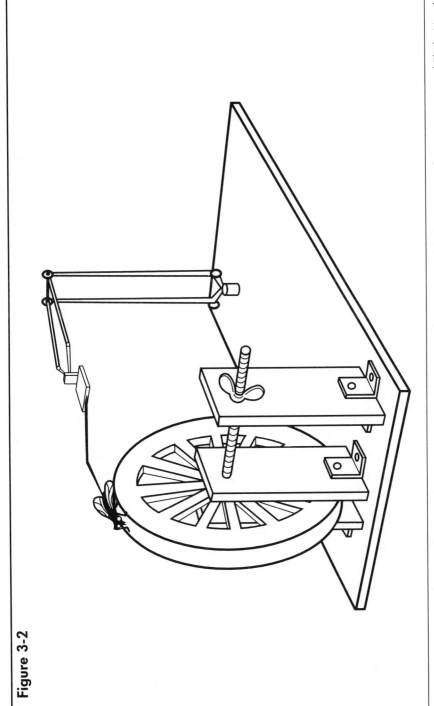

A running wheel for the housefly. From "The Effect of Light, Temperature, and DDT Poisoning in Housefly Locomotion and Flight Muscle Activity" by T. Miller, L. J. Bruner, and T. R. Fukuto, 1971, *Pesticide Biochemistry and Physiology, 1*, p. 484. Copyright 1971 by Academic Press, Inc. Adapted with permission.

confined within the wheel but rather is tethered and allowed to walk on the rim. Photocells record the rotation of the wheel in both designs.

Actograph

Actographs are a family of devices used to measure activity of an animal. Actographs come in many shapes and sizes, typically in the form of a cage. Movements within are recorded using, for instance, mechanical, photoelectric, ultrasonic, acoustical, and video devices. An activity wheel is considered one type of actograph, but generally an actograph measures activity of the subject within its home environment. Depending on the sensitivity of the recording system, it is possible to distinguish several classes of behavior and their neuronal correlates. Rearing, feeding, flying, movement, and direction of movement are the most common behaviors recorded. Like the activity wheel, the principal application of an actograph is in the area of pharmacology and rhythmic behavior. It is used also to study the effect of motivation on behavior, such as the relation between hunger (or thirst) and activity. Of special interest are flight mills, which are actographs designed to study the mechanisms of insect flight. Actographs are useful also in the construction of an ethogram, that is, a catalog of the behavioral repertoire of an animal.

An interesting use of actographs for studies of learning and memory is general activity conditioning. In the most basic example, a CS such as light is paired with a US such as food. After a number of light–food pairings, the animal begins to increase its activity during the light. The "anticipatory" activity associated with stimuli preceding food is certainly familiar to anyone who has observed animals, such as fish, at feeding time. Such activity is easily recorded by an actograph. Although general activity conditioning is feasible, for many invertebrates no studies have appeared in the literature.

Actographs are available for all major groups of invertebrates. See Atkinson, Bailey, and Naylor (1974), Miller (1979), and Dalley and Bailey (1981) for reviews of early devices. An actograph suitable for invertebrate learning experiments has been developed by Brunner and Maldonado (1988), and a computerized infrared monitor for aquatic animals is also available (Kirkpatrick, Schneider, & Pavloski, 1991). Olivo and Thompson (1988) provide an excellent discussion of video digitization techniques for monitoring animal movements. Of special interest for learning experiments is the development of an ultrasonic transmitter for animal tracking and biotelemetry of muscle activity from free-ranging marine animals

(Wolcott & Hines, 1989). This device was designed to measure feeding activity in foraging blue crabs and can easily be adapted for studies of learning. Wind tunnels designed for pheromone research also are available and can be adapted for invertebrate learning research (Jones, Lower, & Howse, 1981; Miller & Roelofs, 1978; Sanders, 1985; Sanders, Lucuik, & Fletcher, 1981). Figure 3-3 depicts an apparatus developed by Zhanna Shuranova of the Institute of Higher Nervous Activity and Neurophysiology in Moscow to measure the activity of crayfish. The animal is placed in a small aquarium, and the turbulence produced by the animal as it moves about is detected by a transducer.

Maze

A *maze* is a seminatural environment consisting of a number of pathways and choice points. Most mazes are separated into a start box, alley segment, choice point, and goal box. The animal is placed in the start box for the beginning of a trial, and the door to the start box is lifted, permitting the animal to run down the alley until it reaches a choice point (for a T maze, the choice point is the cross of the T). If the choice is correct, the animal finds a favorable goal box consisting of a reward such as food or a return to the home cage. If the choice is incorrect, the goal box is unfavorable and is often followed by a punishment such as shock. Maze performance can be quantified in several ways, the most common being the time needed to leave the start box, speed in negotiating the maze, and the number of errors per trial. Some aspect of the locomotor response may also be recorded, such as retracing, the distance traveled down a path, or in the case of multiple-unit mazes, a running sequence of choices. The experiment is usually terminated after some criterion of mastery is reached, such as several consecutive errorless runs through the maze.

There are more than 100 different maze patterns. Generally, a maze may be spatial or temporal. The spatial maze—which is the most common—comes in a variety of designs from simple Y or T shapes to complex designs resembling a New York City subway map. The characteristic that separates the spatial design from the temporal is that the animal can use visual, auditory, olfactory or kinesthetic cues to solve the spatial maze. In contrast, information provided by the senses are not effective in solving a temporal maze. Rather, the solution is thought to involve some type of cognitive processing. In other words, the animal must remember the solution in the absence of cues. In contrast to the large number of spatial

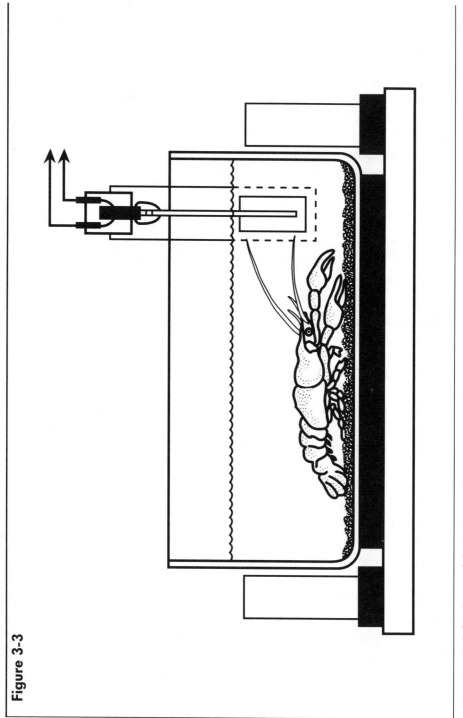

Figure 3-3

An actograph for the crayfish. Courtesy of Zhanna Shuranova.

maze investigations with invertebrates, the ability of invertebrates to solve temporal mazes has not been examined.

The first mazes were quite complex, requiring the animal to make many choices. Discovering how an animal solves such a complex problem became so elusive, however, that simple maze designs were employed. There is now a renewed interest in complex maze designs for vertebrates.

The maze is an extremely useful and versatile apparatus. It is used in many types of experiments involving, for example, foraging behavior, discrimination and choice, probability learning, reversal learning, pattern of reward, and latent learning. By varying the distance from the choice point to the goal box it is possible to manipulate the delay of reinforcement. A major drawback of the maze is that the animal is usually handled between trials, and in the case of invertebrates, must often be induced by the experimenter to leave the start box or alleyway.

A second problem is that the length and size of the start box, alleyway, and goal box segments of invertebrate mazes (and runways) are not standardized. Nor is the distance between the choice point and the goal boxes standardized. Each of these distances contributes to performance. The failure to standardize these distances in invertebrate mazes contributes to the difficulty of comparing the performance of invertebrates across experiments and across species. For example, the superior T-maze performance of ants compared to earthworms may be due, as many suggest, to the advances of the arthropod nervous system over their annelid relatives. Alternatively, it would not be surprising to discover that performance is better because the ant reaches the goal box more quickly than an earthworm, thereby minimizing the delay of reinforcement.

A third problem that also is unique to invertebrates is finding a suitable reward. In most invertebrate maze (and runway) experiments, the subject's motivation to run is not the search for food but rather the escape from a hostile environment. Because many invertebrates can survive long periods without food or water, one approach is to equate the maze with the foraging situation by connecting it to the nest or individual home container. A second and more general solution is to use a preferred food.

Another limitation of the maze is that it is difficult to automate completely and is therefore labor-intensive. There is also the problem of separating classical from instrumental effects in maze performance. Performance, for example, can easily be interpreted not as "behavior controlled by its consequences"—that is, instrumental conditioning—but rather as the classical conditioning of approach responses to cues preceding food.

Mazes are available for all of the major groups of invertebrates. Arguably, the maze is the apparatus most commonly used for the study of invertebrate learning. The vast majority of the devices, however, are of the simple T and Y variety. The most complex designs are found in studies of ant learning. Mazes are available for ants (Schneirla, 1933; Stratton & Coleman, 1973; Vowles, 1964), *Aplysia* (Preston & Lee, 1973), bees (Menzel & Erber, 1972), crabs (Datta, Milstein, & Bitterman, 1960; Yerkes, 1902), crayfish (Capretta & Rea, 1967), crickets (Wilson & Hoy, 1968), earthworms (Datta, 1962; Swartz, 1929), fruit flies (Drudge & Platt, 1979; Dudai, 1977), grain beetles (Alloway, 1972), lobsters (Schöne, 1961), roaches (Longo, 1964), and planarians (Best & Rubinstein, 1962; Corning, 1964). Darchen (1964) and McConnell (1967a, 1967b) discuss some of the factors that contribute to successful maze performance in cockroaches and planarians respectively.

Figure 3-4 shows the complex maze developed by Schneirla (1929) in his studies of ant learning. The start box is connected to the nest, and by a suitable arrangement of doors, different maze patterns are created. Handling is eliminated in this design by patiently waiting for an individual ant—usually identified by numbered tags or paint—to appear at the maze entrance, at which time it is admitted to the start box.

Figure 3-5 depicts an interesting version of a Y maze developed by Wilson and Hoy (1968) for studies of the optomotor response in the milkweed bug. The animal is tethered and allowed to grasp the maze, which is constructed from Styrofoam. An interesting feature is that the animal is not handled between trials, and complete automation is possible.

Runway

A *runway* is a maze without blind alleys. Like the maze, a runway consists of start box, alleyway, and goal box segments. A trial begins by placing the animal in the start box. The door is opened, and the animal is allowed to run down the alley into the goal box or until a period of time has elapsed, at which point the animal is returned to the start box for a new trial. Runway performance can be measured in several ways, such as the time to leave the start box (starting time) and the time to traverse the alley (running time). The runway can easily be adapted for choice experiments by either changing some characteristic of the alley (or goal box) such as color or substrate or by using two runways that share a start box.

The runway is most commonly used to investigate parameters of reward such as frequency, probability, amount, quality, and delay. Escape

Figure 3-4

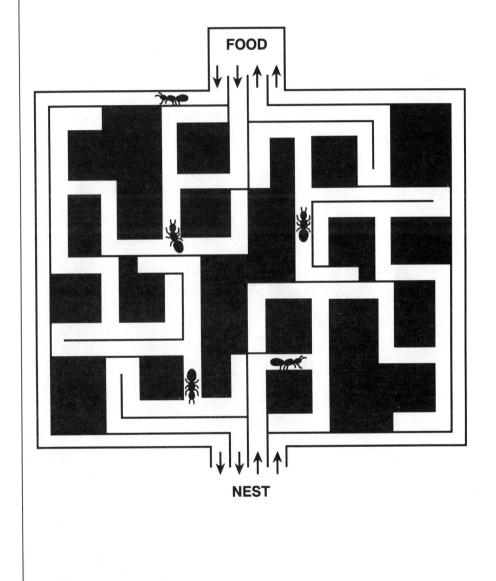

A complex maze for the ant. From "Motivation and Efficiency in Ant Learning" by T. C. Schneirla, 1933, *Journal of Comparative Psychology, 15,* p. 245. Copyright 1933 by the American Psychological Association.

Figure 3-5

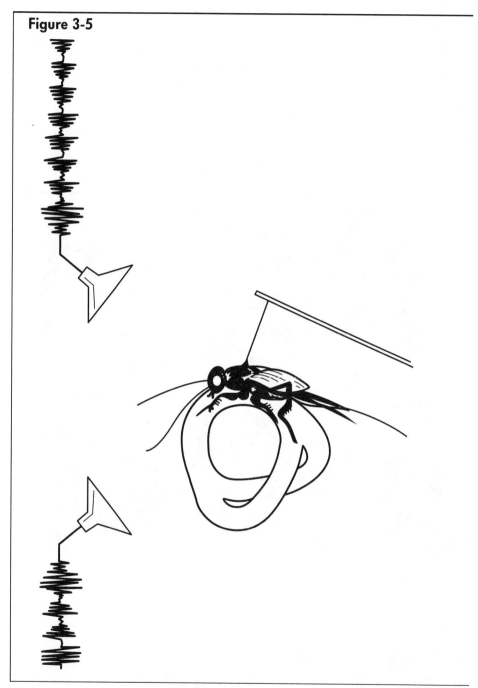

A Y maze for the milkweed bug. From "Optomotor Reaction, Locomotory Bias, and Reactive Inhibition in the Milkweed Bug *Oncopeltus* and the Beetle *Zophobas*" by D. M. Wilson and R. R. Hoy, 1968, *Zeitschrift für Vergleichende Physiologie, 58*, p. 138. Copyright 1968 by Springer-Verlag. Adapted with permission.

and avoidance procedures are also possible. Compared to the maze, the runway has seldom been used in the study of invertebrate learning. An interesting use of the runway is the technique developed by Logan (1960) to stimulate free-operant schedules of reward such as simple ratio and interval schedules. Considering that there are few invertebrate free-operant lever-press devices (see item 5), it would be of great interest to apply Logan's techniques to invertebrates. A limitation of the apparatus is that the animal must be handled between trials. Handling can be eliminated by using interchangeable start and goal boxes. Performance is better, however, if the start and goal boxes remain distinct.

Runways are available for ants (Abramson, Miler, & Mann, 1982), earthworms (Reynierse & Ratner, 1964), marine worms (Copeland, 1930; Evans, 1966), and roaches (Longo, 1970). Kirk and Thompson (1967) provide information on the factors that influence the runway performance of earthworms. Figure 3-6 shows a runway designed for roaches (Longo, 1970). The start and goal boxes are interchangeable, and the motivation to run is the opportunity to consume a preferred food. The position of the subject is monitored by photocells and detectors.

Lever-Press Box

A *lever-press box* (also known as a *Skinner box*, *operant chamber*, or *automated problem box*) is an environment in which an animal learns to manipulate some device to obtain a reward. The operant chamber has become a standard tool for the study of vertebrate behavior. The apparatus is fully automated and can accommodate a wide range of experiments, including those concerned with discrimination and generalization, timing, schedule effects, choice behavior, foraging, escape, punishment, avoidance, conditioned suppression, behavioral toxicology, and psychophysics. In addition to versatility, an advantage of the apparatus is that it facilitates species comparison because it minimizes species-typical behavior patterns. The latter strength is also a limitation for those interested in such behavior patterns. Another limitation is that the ease of making the manipulative response may obscure the effect of some training variable that has been found to be important in runway and maze situations, that is, in situations requiring a distinctive response. The operant chamber is also not efficient for studies of avoidance. When applied to invertebrates, the technique is often difficult to use, and performance can be highly variable.

The measure of performance is usually the rate of response, but other measures are possible, such as latency and—with two manipulan-

Figure 3-6

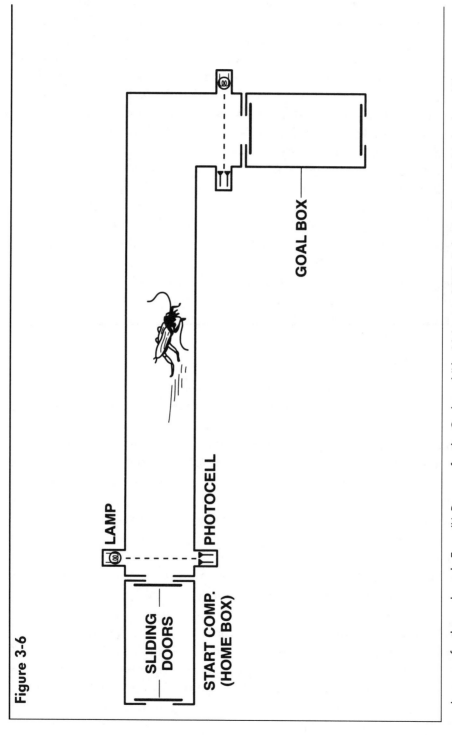

A runway for the cockroach. From "A Runway for the Cockroach" by N. Longo, 1970, *Behavior Research Methods and Instrumentation*, 2, p. 118. Copyright 1970 by Psychonomic Society Publications. Adapted with permission.

Figure 3-7

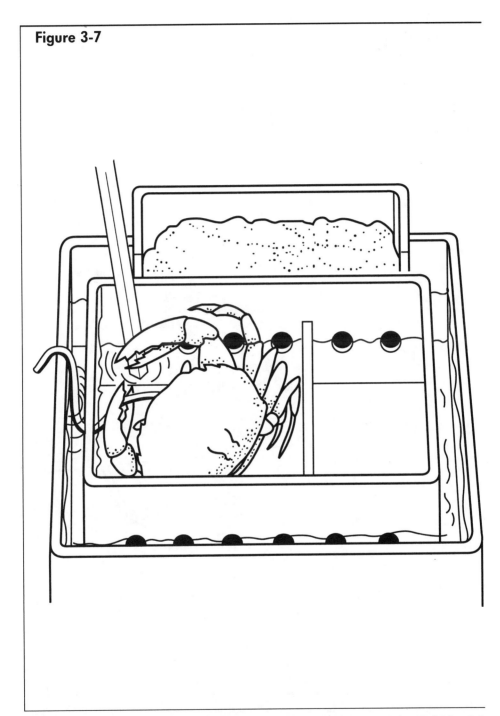

A lever-press situation for the crab. From "Lever Press Conditioning in the Crab" by C. I. Abramson and R. D. Feinman, 1990, *Physiology & Behavior, 11*, p. 268. Copyright 1990 by Pergamon Press. Adapted with permission.

Figure 3-8

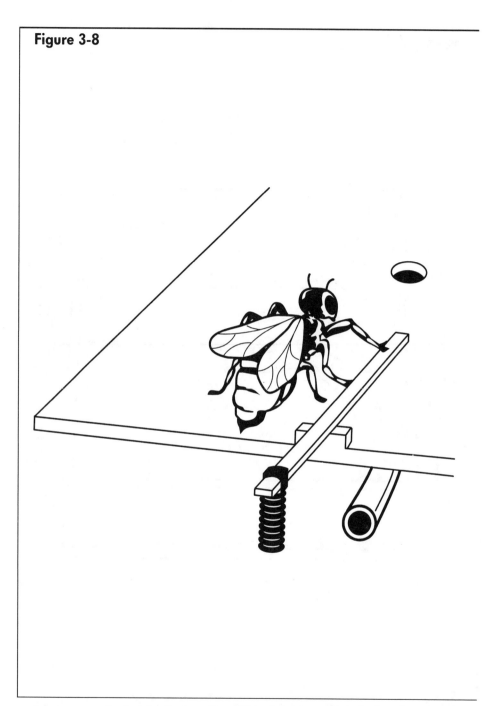

A lever-press situation for the bee. From "Discrimination with Light Stimuli and a Lever-Pressing Response in *Melipona rufiventris*" by I. Pessotti, 1972, *Journal of Apicultural Research, 11*, p. 90. Copyright 1972 by the International Bee Research Association. Adapted with permission.

Figure 3-9

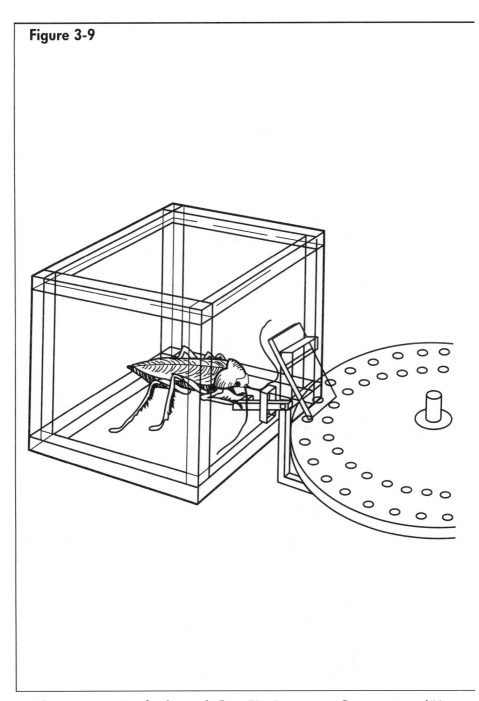

A lever-press situation for the roach. From "An Apparatus to Demonstrate and Measure Operant Behavior of Arthropoda" by D. O. Rubadeau and K. A. Conrad, 1963, *Journal of the Experimental Analysis of Behavior, 6,* p. 429. Copyright 1963 by the Society for the Experimental Analysis of Behavior, Inc. Adapted with permission.

Figure 3-10

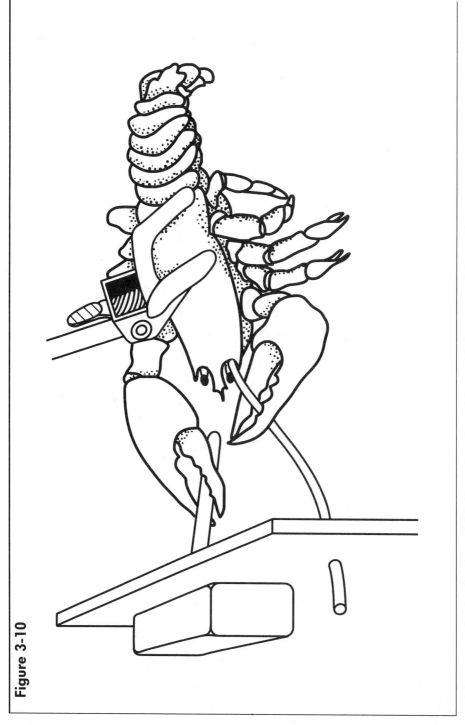

A lever-press situation for the crayfish. Figure based on slide provided courtesy of Gene Olson.

Figure 3-11

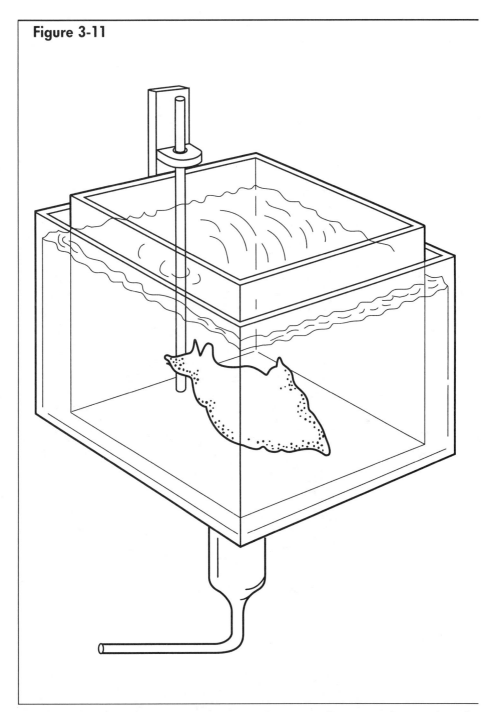

A lever-press situation for *Aplysia*. From "Cooling as Reinforcing Stimulus in *Aplysia*" by P. Downey and B. Jahan-Parwar, 1972, *American Zoologist, 12*, p. 508. Copyright 1972 by the American Society of Zoologists. Adapted with permission.

Figure 3-12

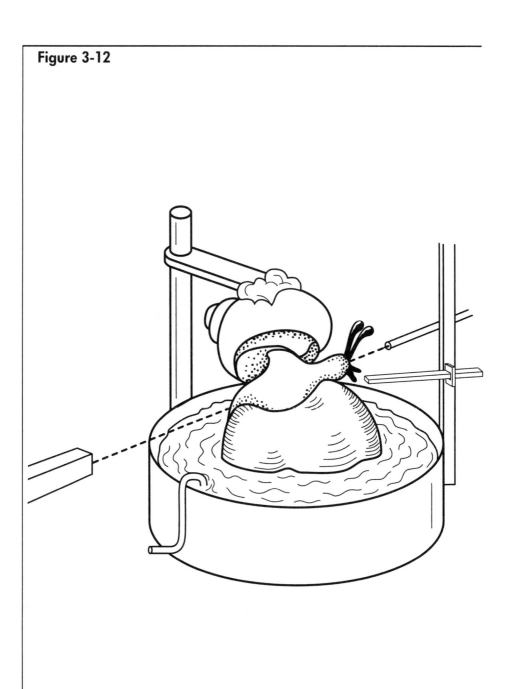

A lever-press situation for the snail. From "Self-Stimulation in Snails" by P. V. Balaban and R. Chase, 1989, *Neuroscience Research Communications, 4*, p. 141. Copyright 1989 by John Wiley & Sons, Ltd. Adapted with permission.

dums present—choice. The first lever-press situation for invertebrates was described by Rubadeau and Conrad (1963) for the roach. The lever was fashioned out of the needle from a volt meter. Downey and Jahan-Parwar (1972) describe a device for *Aplysia*. Operant boxes are also available for bees (Pessotti & Lignelli-Otero, 1981), crabs (Abramson & Feinman, 1990a), crayfish (Olson & Strandberg, 1979), and snails (Balaban & Chase, 1989). Several figures depict varying types of operant chambers developed for crab (Figure 3-7), bee, (Figure 3-8), roach (Figure 3-9), crayfish (Figure 3-10), *Aplysia* (Figure 3-11), and snail (Figure 3-12).

Shuttle Box

A *shuttle box* (also known as a *choice chamber* or *double chamber*) is a chamber of two or more compartments that requires an organism to move from one compartment to another. Some designs are circular in shape rather than rectangular. For vertebrates, the shuttle box has become extremely useful for the study of escape, avoidance, and punishment. Its use by psychologists for the study of invertebrate behavior is rather limited to studies of escape, punishment, and some avoidance studies. Behavioral biologists, however, have made substantial use of choice chambers in their studies of how the behaviors of invertebrates are influenced by such independent variables as temperature, humidity, and illumination. The primary differences between shuttle boxes and choice chambers are the independent variables. Fraenkel and Gunn (1961) provide many examples of the use of choice chambers with invertebrates.

A limitation of the device is that, at least for vertebrates, animals are reluctant to reenter a compartment in which they received aversive stimulation. This has profound effects on the efficiency of avoidance behavior as measured in a two-way shuttle box. Whether such behavior occurs with invertebrates is unknown. The reluctance is reduced by using a circular shuttle box in which the animal always moves in a clockwise or counter-clockwise direction or a one-way shuttle box in which the animal always goes in one direction. The primary data in studies employing shuttle boxes are of latency: the time taken to move from one compartment to another following the onset of the conditioned or unconditioned stimulus. The amount of time spent in each compartment is also an important measure of performance.

Shuttle boxes are available for ants (Abramson et al., 1982), confined bees (Abramson, 1986), and free-flying bees (Lee & Bitterman, 1990a), crabs (Fernandez-Duque, Valeggia, & Maldonado, 1992; Karas, 1962),

Figure 3-13

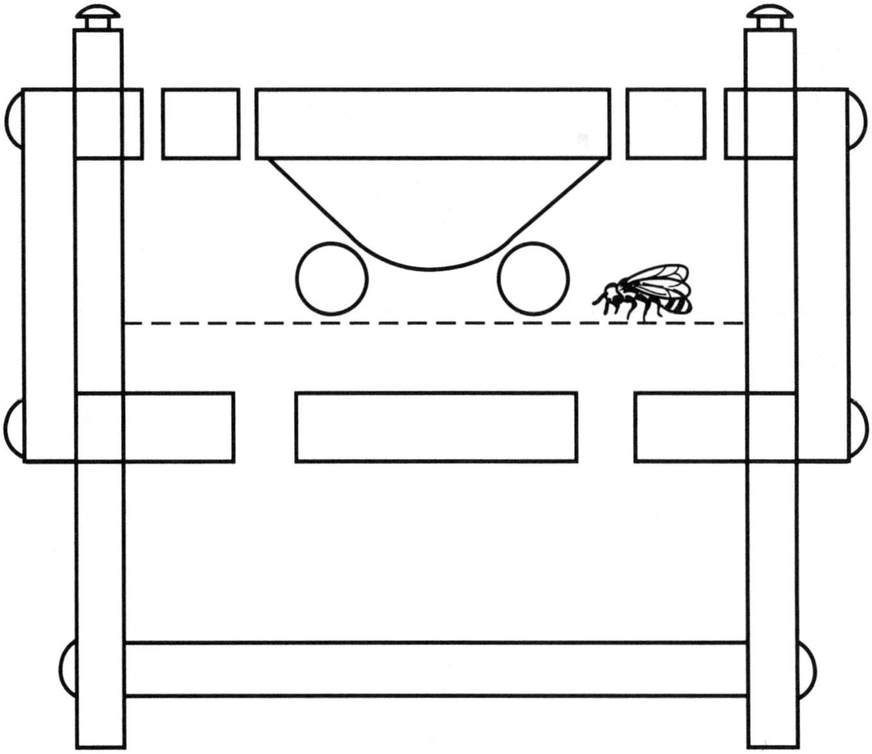

A shuttle box for the bee. From "Aversive Conditioning in Honeybees (*Apis mellifera*)" by C. I. Abramson, 1986, *Journal of Comparative Psychology, 100*, p. 109. Copyright 1986 by the American Psychological Association.

Figure 3-14

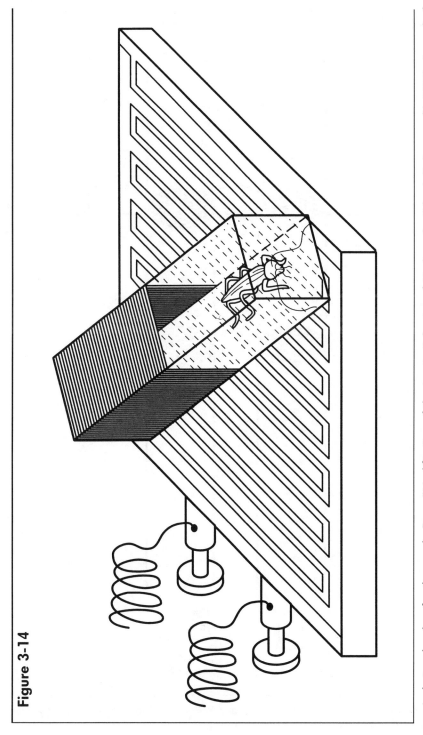

A choice chamber for the roach. From "Modification of the Innate Behavior of Cockroaches" by J. S. Szymanski, 1912, *Journal of Animal Behavior, 2,* p. 81. Copyright by *Journal of Animal Behavior.*

Figure 3-15

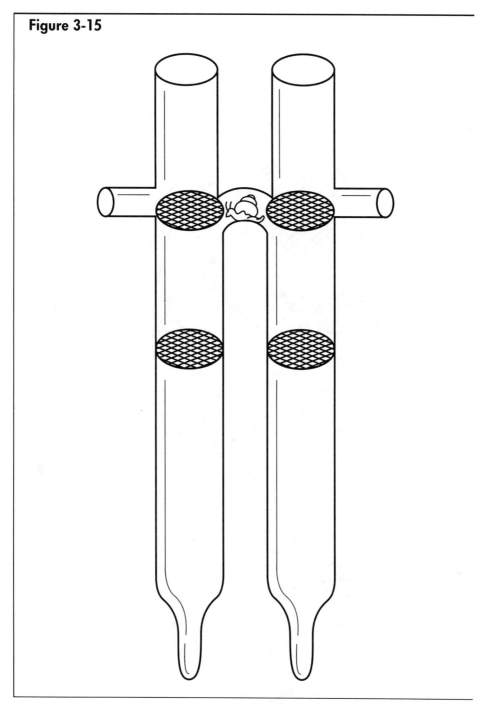

A shuttle box for the snail. From "Conditioned Reflexes in Snails *Physa acuta*" by V. A. Sokolov, 1959, *Vestnik Leningrad University, 9,* p. 82.

Figure 3-16

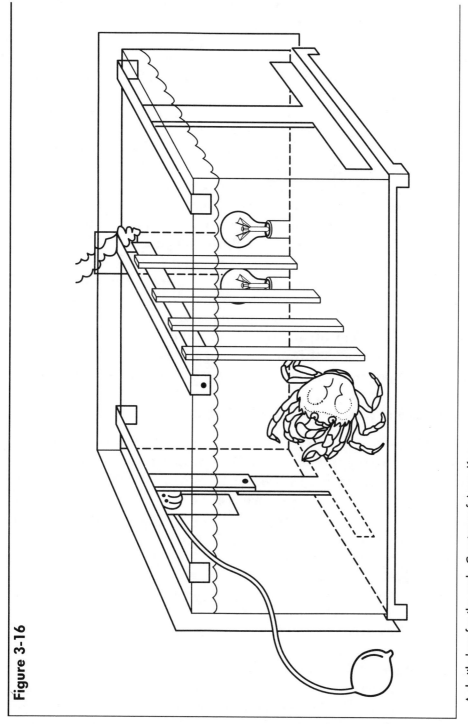

A shuttle box for the crab. Courtesy of Anna Karas.

crayfish (Taylor, 1971), horseshoe crabs (Makous, 1969), houseflies (Leeming & Little, 1977), roaches (Brown & Stroup, 1988; Szymanski, 1912), and snails (Sokolov, 1959). Figure 3-13 depicts a shuttle box designed for the honeybee (Abramson, 1986). The two compartments are divided by a hurdle that the bee must crawl under. Photocells determine the location of the bee.

Szymanski's (1912) apparatus for the study of learning in the cockroach is diagrammed in Figure 3-14. One half of a glass rectangular box is covered black. The box is inverted and rests upon a shock grid. The roach is introduced on the lighted side; it immediately scurries to the dark side, at which time it receives a shock that drives it back into the light. After several dark–shock pairings, the animal remains in the light.

Figure 3-15 depicts a Russian technique developed by Sokolov (1959) for training snails. The animal is confined in one section of a tube. The animal shuttles to a second tube by crawling through a U-shaped bridge.

Figure 3-16 depicts a Russian shuttle box used to study instrumental reward conditioning in the green crab (Karas, 1962). The apparatus is unique in that various discriminative stimuli such as light, bubbling water, and substrate movement are used to signal the availability of food. The movement of the animal from one compartment to the other is automatically recorded whenever the crab passes through the center gate.

Conditioning of Food Aversion and Food Preferences

Conditioning of food aversion and food preferences (also known as *taste aversion learning*, *food aversion learning*, or *food preference conditioning*) is a classical conditioning technique in which the CS is taste and the US is poison. After eating or drinking, the animal is made ill by injection, radiation, or CO_2 narcosis. Following recovery, the animal is given a choice between the substance that served as the CS and a control substance. In the conditioning of food preferences, an odor or taste CS is paired with a palatable or favored food. The measure of learning is the relative amounts of the substance consumed or, alternatively, the amount of time spent in the presence of the CS. The advantage of the procedure is that it produces very robust and quite long-lasting changes in consummatory behavior. In addition, it is well suited for quantitative physiological and biochemical analysis of the learning process.

The principal disadvantage of the technique is similar to that to be discussed later for conditioning of contraction: the problem of interpretation. As will be discussed in chapter 5, in what sense is taste a CS? If taste aversion is not an example of classical conditioning, is it (a) an example of discriminative punishment in which a response—eating—is punished by illness; (b) an example, as Solomon (1977) suggests, of an association between two unconditioned stimuli; or (c) a new conditioning procedure?

Taste aversion has been demonstrated in caterpillars, (Dethier, 1980), crabs (Wight, Francis, & Eldridge, 1990), locust (Bernays & Lee, 1988), slugs (Gelperin, 1975), and the gastropod mollusc *Pleurobranchaea* (Mpitsos & Collins, 1975). Food preferences have been established in the snail (Audesirk, Alexander, Audesirk, & Moyer, 1982; Alexander, Audesirk, & Audesirk, 1984) and several phytophagous insects (for a review of the insect studies, see Papaj & Prokopy, 1989). Figure 3-17 shows a popular device for testing the odor preferences of slugs. Following training, an animal is placed in the center or "neutral zone" of a plastic dish measuring 14 cm in diameter. On either side of the neutral zone are the test and control odors, respectively.

Proboscis Conditioning

Proboscis conditioning (also known as *olfactory conditioning* or *proboscis extension reflex conditioning*) is an invertebrate classical conditioning technique in which an olfactory stimulus (CS) is paired with a sucrose feeding (US). After several CS–US pairings, extension of the proboscis—a tube through which liquids are sucked into the mouth—is evoked by the CS, which at the outset of training is ineffective. The primary data consist of the presence or absence of a proboscis extension following presentation of the relevant stimuli. By employing video technology or physiological recording, fine-grain analysis of the proboscis response can readily be achieved. First developed by Frings (1944) to determine sensory thresholds in blow flies, the proboscis technique has become the most popular technique for classical conditioning in the bee.

The technique permits excellent control of intertrial interval, CS–US interval, stimulus intensity, and stimulus duration. The importance of these and other training variables is easily investigated. It is also possible to compare directly the performance of free-flying foragers—trained using some variant of the von Frisch technique (see item 10)—with those trained under restrained conditions. This is the only invertebrate in which such a comparison is possible. In addition to behavioral experiments, the proboscis technique is well suited also for quantitative physiological and biochemical analysis of the learning process.

Figure 3-17

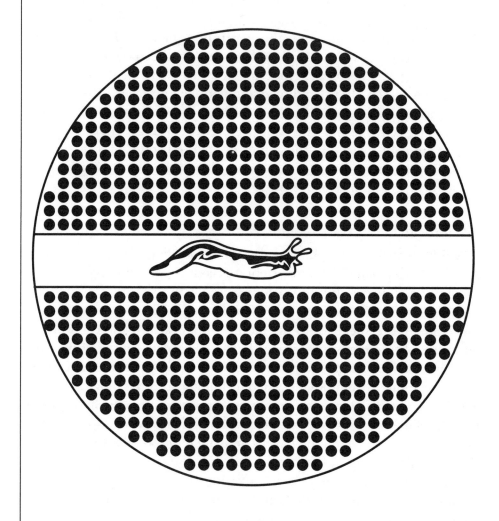

A food-odor preference device. From "One-Trial Learning Modifies Food Odor Preferences of a Terrestrial Mollusc" by C. L. Sahley, A. Gelperin, and J. W. Rudy, 1981, *Proceedings of the National Academy of Sciences (USA), 78*, p. 641. Copyright 1981 by C. L. Sahley. Adapted with permission.

The principal limitation of the technique is that the physical condition of the restrained animal deteriorates after 24 hours. There is also a problem of satiation with repeated sucrose feedings; after that point, sucrose feeding will no longer serve as an effective US. Both factors combine to limit the length of experiments. Satiation can be controlled somewhat by using a minute amount of sucrose or reducing the feeding time. Manipulating the amount and duration of the US is, however, known to influence the performance of vertebrates in classical conditioning experiments. Satiation can be controlled also by surgically removing the abdomen. This operation allows the sucrose to flow out of the animal. Such a radical solution is not without complications. The proboscis situation is also difficult to automate and CSs other than odors are seldom effective.

The problem of automation is especially important in the proboscis technique. Vargo, Holliday, and Hirsch (1983) have identified four ways in which the manual presentation of stimuli may influence the results. Experimenters have been found to differ in (a) the CS and US durations, (b) the location of stimulation, (c) the amount of disturbance the animal goes through during the conditioning procedure, and (d) the consistency of technique from trial to trial. Their solution to the problem is to present stimuli on a kymograph drum.

Proboscis conditioning situations are available for bees using positive (Bitterman, Menzel, Fietz, & Schäfer (1983), and aversive unconditioned stimuli (Smith, Abramson, & Tobin, 1991), for blowflies (Akahane & Amakawa, 1983; Maes & Bijpost, 1979; Nelson, 1971), and houseflies (Fukushi, 1979).

Figure 3-18 presents a proboscis conditioning situation for the bee. The animal is confined to a tube that is placed in front of an exhaust fan to remove unwanted olfactory cues. The CS and US are presented manually.

Horridge Procedure

The *Horridge procedure* (also known as *leg position learning* or *position learning*) is an instrumental conditioning technique for invertebrates in which movement of an appendage beyond a specified position is reinforced. This paradigm was one of the first to suggest that invertebrates can be used to investigate the neuronal basis of learning and memory. In the original version, two animals are suspended above an electrified solution. The experimental animal receives a shock whenever its leg dips below a certain position. A control animal is connected (or yoked) to the experimental animal such that it too receives a shock whenever the experimental

Figure 3-18

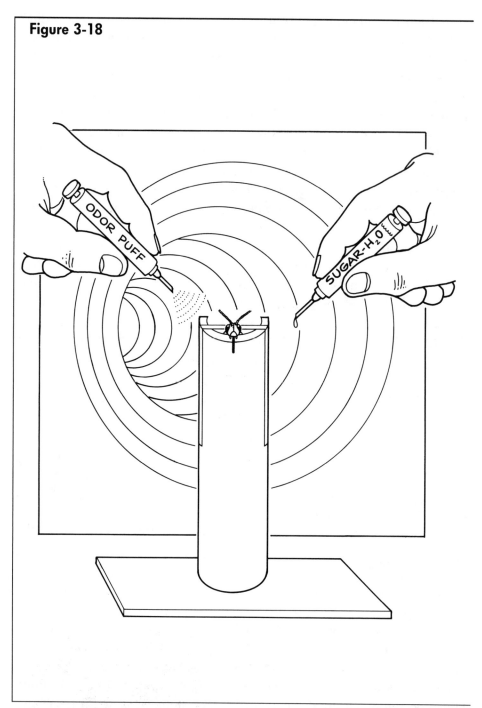

A proboscis conditioning situation for honeybees. From *Invertebrate Learning: A Laboratory Manual and Source Book* (p. 39) by C. I. Abramson, 1990, Washington, DC: American Psychological Association. Copyright 1990 by the American Psychological Association.

animal does but independently of its own leg position. Learning is inferred by a difference in the number of shocks received between experimental and yoked animals when both are free to receive shocks contingent upon leg position. The basic data are the number of shocks received. The technique need not be restricted to leg position learning. Versions are available to condition eye position in crabs, claw position in crayfish, optic tentacle position in snails, and head waving in molluscs.

The technique was quite popular during the 1960s and 1970s but, with the exception of head waving in *Aplysia*, was used fairly seldom in recent years. There are limitations. First, the results cannot always be replicated, especially with individual animals, and not all experimental animals learn. Second, there is a problem of interpretation. The leg lift paradigm seems simple enough; it is difficult to determine, however, if such learning is based on escape from aversive stimulation, the result of a punishment contingency, or due to avoidance behavior. A third problem is that shock produces global effects that interfere with the interpretation of nerve recordings and produce unconditioned movements of the leg. Fourth, the results seem to depend not as much on the contingencies of the experiment as on the position of the leg at the start of training. An improved version for the roach has recently appeared (C. L. Harris, 1991). The differences between experimental and control animals were not large during training or testing. However, significant effects did occur during reversal learning when the control animals could control the number of shocks.

Recently, however, Forman (1984) developed a design for leg position learning in the locust. Rather than simply requiring the animal to raise a leg to terminate a series of shocks, the locust is required to maintain a particular range of leg movement arbitrarily selected by the experimenter. After a few minutes of training, the animal learns to shift its leg position to an angle that terminates aversive heat to the head or, alternatively, produces access to food. The task can be made more complex by narrowing the range or by shifting the range. It is also possible to train a cold locust to manipulate its leg to turn on the heat lamp. In other words, it can be trained to thermoregulate.

The Forman technique is a significant advance because the response and reinforcer are arbitrary, and learning can be identified in an individual animal. If you are considering the Horridge paradigm, I suggest you adapt the Forman version. It should also be noted that although the Horridge paradigm is designed as an instrumental training situation, one can imagine that it can easily be adapted for classical flexion conditioning.

To do so, remove the response–reinforcer contingency and add a CS. Following a number of CS–shock pairings, you would hope to see the leg (or any another appendage) flex.

Leg position techniques are available for crabs (Dunn & Barnes, 1981a, 1981b; Hoyle, 1976), fruit flies (Booker & Quinn, 1981), locust (Forman, 1984), and roaches (Horridge, 1962). It has been used also on the spinal frog preparation (Farel & Buerger, 1972). The Horridge procedure can be used to train appendages other than the leg, such as eye withdrawal in crabs (Abramson & Feinman, 1987), tentacle movement in snails (Christoffersen, Frederiksen, Johansen, Kristensen, & Simonsen, 1981), body orientation in fruit flies (Mariath, 1985), and claw movement in crayfish (Stafstrom & Gerstein, 1977). The procedure can also be generalized to include those situations in which the innate behavior of the animal is punished, as, for example, when a roach is punished for entering a normally attractive dark compartment (Szymanski, 1912). An interesting version of this technique has been reported by Carrega and Huber (1985), who trained roaches to refrain from entering a compartment that contains an attractive odor. Figures 3-19 through 3-23 depict several types of Horridge designs developed for crabs (Figures 3-19 and 3-20), crayfish (Figure 3-21), roach (Figure 3-22), and fly (Figure 3-23).

Free-Flying Procedure

The *free-flying procedure* (also known as the *von Frisch technique*) is an instrumental training technique for honeybees in which trained animals shuttle back and forth from the laboratory colony to the experimental arena, where they take sucrose from targets distinguished by color, odor, or position. Developed by Karl von Frisch (1914), the free-flying procedure has become one of the most important in the analysis of invertebrate learning. The principal advantages are versatility and convenience. Some rather sophisticated experiments have been performed using this procedure, including overshadowing, potentiation, and within-compound associations; dimensional shift in choice problems; the overlearning–extinction effect and its dependence on magnitude of reinforcement; contrast effects; operant conditioning of lever press and head dipping; punishment and signaled avoidance. Other advantages are the simplicity of the apparatus, the naturalness of the situation, and the ability to do experiments lasting several hours or days. The latter advantage is perhaps unique to the bee in the invertebrate learning literature because unlike other invertebrates, the bee does not satiate. After a feeding, the bee returns to the nest and unloads its booty before alighting for the return trip to the test arena.

Versatility, sophistication, and convenience of the technique have no doubt contributed to the fact that we know more about the learning of bees than any other invertebrate. Nevertheless, the technique has limitations. The most serious is the fact that it is difficult to control such training variables as intertrial interval and stimulus durations. These variables are determined by the bee. It is also not possible to give the bee an unrewarded trial. This problem can be circumvented by delaying the sucrose feeding or temporarily substituting water for sucrose. Other limitations are the difficulty of employing physiological and biochemical techniques, the inability to use certain types of experimental designs, the unavailability of bees during colder months, the influence of climatic conditions during an experiment, and the effort in maintaining bee hives. You should also be aware that like all of the techniques described in this chapter, the underlying behavioral mechanisms of the free-flying procedure are difficult to determine. The procedure is instrumental in character, but like the maze and runway, it is easy to conceive that the controlling variable is the classical conditioning of approach responses to the target.

The performance of free-flying animals can be quantified in several ways. The most common ways are the number of times the bee contacts the target(s), choice, time spent flying over a target (hover time), and the time needed to return to the target following a visit to the colony (return time). An example of the free-flying technique can be found in Couvillion and Bitterman (1980). Figure 3-24 depicts an overview of the free-flying technique, in which the bee is confronted with a choice between two targets, one with sucrose (S+) and the other with water (S−). The bowls seen in the left-hand portion of the figure are used to remove any scent the animal may have left on the targets.

An analogous technique has recently been developed by Fukushi (1983, 1985) for training houseflies and blow flies, Bernays and Wrubel (1985) for the grasshopper, and Pantaleâo and Morato (1989) for the Mediterranean fruit fly. An individual animal is trained to feed on a scented sucrose droplet (or colored droplet) and then, while foraging, given a choice between the training stimulus and a control. A similar situation is available for the crab and shrimp (Mikhailoff, 1923) which can be taught to discriminate between two colored cylinders, one of which was previously associated with food during foraging trips.

Figure 3-19

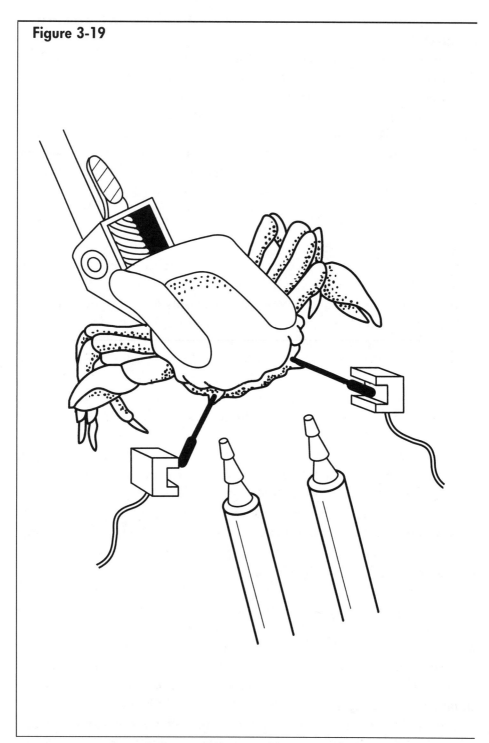

A Horridge eye withdrawal preparation for the crab.

Figure 3-20

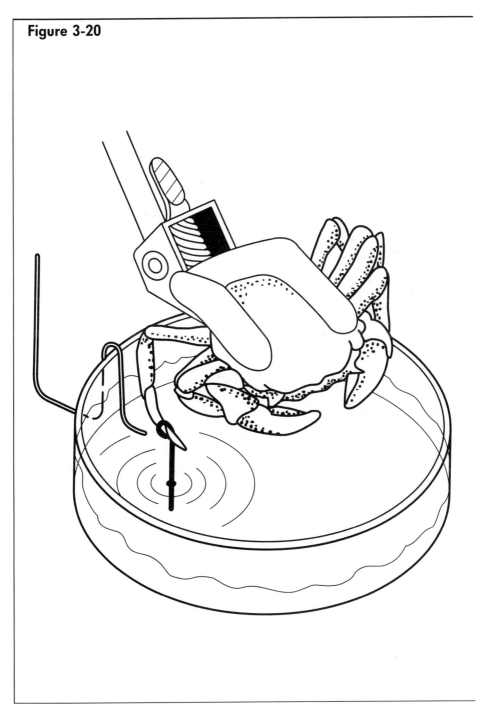

A Horridge leg lift preparation for the crab. From "Learning of Leg Position by the Ghost Crab *Ocypode ceratophthalma*" by G. Hoyle, 1976, *Behavioral Biology, 18,* p. 149. Copyright 1976 by Academic Press, Inc. Adapted with permission.

Figure 3-21

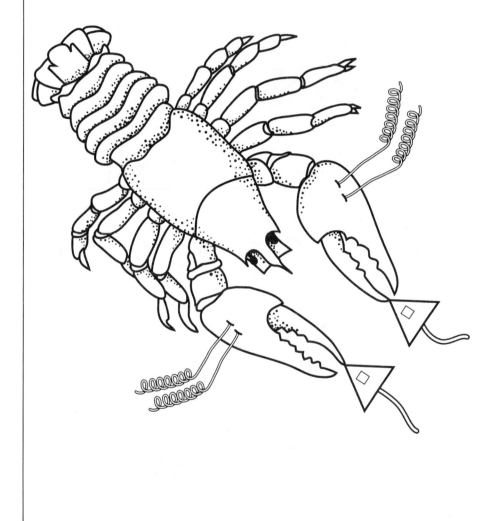

A Horridge claw preparation for the crayfish. From "A Paradigm for Position Learning in the Crayfish Claw" by C. E. Stafstrom and G. L. Gerstein, 1977, *Brain Research, 134,* p. 186. Copyright 1977 by Elsevier/North Holland Biomedical Press. Adapted with permission.

Figure 3-22

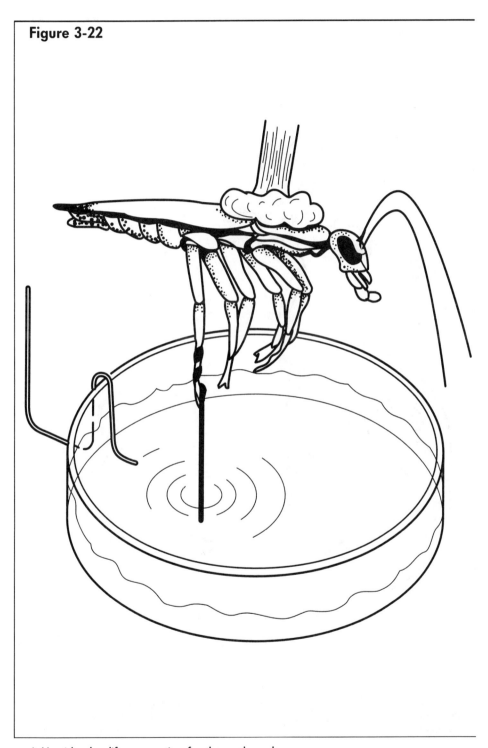

A Horridge leg lift preparation for the cockroach.

Figure 3-23

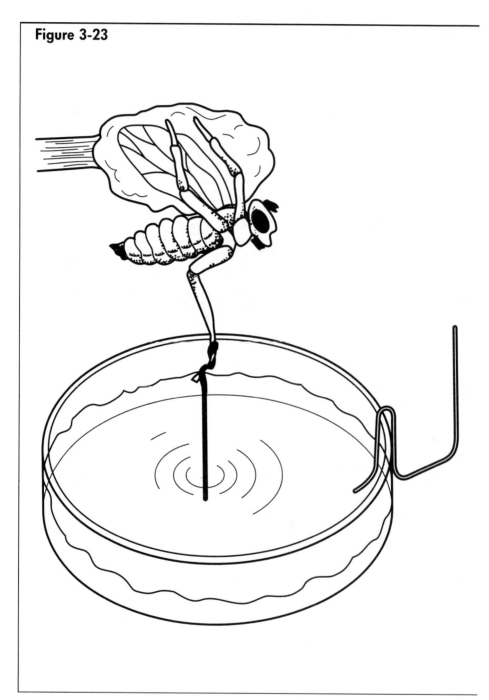

A Horridge leg lift preparation for the fly. From "Conditioning of Leg Position in Normal and Mutant *Drosophila*" by R. Booker and W. G. Quinn, 1981, *Proceedings of the National Academy of Sciences (USA), 78,* p. 3941. Copyright 1981 by W. G. Quinn. Adapted with permission.

Figure 3-24

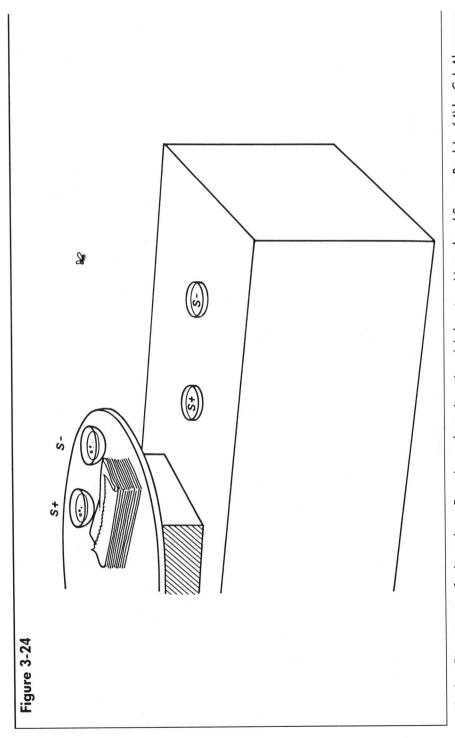

The free-flying situation for honeybees. From *Invertebrate Learning: A Laboratory Manual and Source Book* (p. 64) by C. I. Abramson, 1990, Washington, DC: American Psychological Association. Copyright 1990 by the American Psychological Association.

Countercurrent Apparatus

The *countercurrent apparatus* (also known as the Drosophila *olfactory conditioning situation*) is an olfactory classical conditioning device for training large numbers of fruit flies simultaneously. The technique is used extensively to look at a number of training variables such as (a) the number of training trials, (b) type of conditioned and unconditioned stimuli, (c) motivational variables, (d) retention, and (e) strain of animal. Its primary function, however, is to isolate genetic mutants.

During training, groups of flies are exposed to two sequential odors. One odor is paired with a rewarding event, and the other serves as a nonrewarded control. Following several training experiences with both odors, the animals are given an unreinforced choice test in a T maze. One arm of the maze contains the rewarded odor and the other arm, the control odor. The basic data consist of the fraction of flies choosing the rewarded odor minus the fraction choosing the nonrewarded odor. An index of 0 indicated no learning. In other words, the T-maze choice test revealed that the flies distributed themselves equally in both arms of the maze. An index of 1 indicates that all flies were found in the arm containing the previously rewarded odor, indicating learning; a value of −1 indicates that all flies were found in the arm containing the control odor.

The technique is rather easy to use, and, unlike other devices and techniques we have discussed, it is designed to identify learning in a population of animals rather than in an individual animal. The development of this technique permits the rapid isolation of mutants with altered abilities to learn. Sophisticated biochemical and genetic methodology can then be used to analyze these mutations. Also, the device is easy to use.

There are some limitations of the device. First, the training situation does not ensure that individual animals are exposed to the relevant stimuli. Second, it is not easy to manipulate parametrically the training variables known to influence classical conditioning. Third, the animals have to be transferred from the training to the testing apparatus. Fourth, performance is adversely affected by the influence of phototaxis. In the original procedure, phototaxis was used to stimulate the flies to enter the odor tubes. Fifth, although the device was originally designed for classical conditioning, it became apparent that the underlying behavioral mechanism is based upon instrumental conditioning. These limitations have been addressed in an improved version that employs a better method of presenting the CS and US and eliminates the need to transfer the animals (Tully & Quinn, 1985).

An example of the technique is available from Quinn, Harris, and Benzer (1974) for aversive reinforcement and Tempel, Bonini, Dawson, and Quinn (1983) for positive reinforcement. Figure 3-25 depicts the improved version of the Benzer (1967) apparatus (Tully & Quinn, 1985). A group of flies are placed in a tube containing a shock grid. An improved odor delivery system ensures that the animals receive the conditioned stimuli and eliminates the need for phototaxis. Testing is carried out in a built-in T maze, which eliminates the need for transferring the animals to a separate testing apparatus. The training situation is depicted in Figure 3-25. An enlargement of the testing portion of the apparatus is shown in Figure 3-26.

Classical Conditioning of Contraction

Classical conditioning of contraction (also known as *withdrawal conditioning*) is a family of techniques in which the conditioned response is a contraction of the whole animal or an isolated portion of the animal. Examples of the latter include the conditioning of siphon withdrawal in *Aplysia* (Carew, Walters, & Kandel, 1981) and eye withdrawal in the crab (Abramson & Feinman, 1988). The conditioning of withdrawal is perhaps the most general of the classical conditioning techniques, having been applied to various species of flatworms, annelids, molluscs, and crustaceans. The withdrawal response is measured in several ways, the most common being the probability or amplitude of the withdrawal. The time taken to reextend following a withdrawal is also informative.

The principal advantage of the technique is its ease of use. The withdrawal response is easy to observe and is a natural response for soft-bodied animals and isolated portions of hard-bodied invertebrates. In earlier applications, stimulus presentation and response recording were performed manually. Today, it is quite common to see automated versions in which transducers or photocells record responses and computers present stimuli and dispense training parameters automatically. The technique is also conducive to parametric manipulations of training variables. This advantage is shared with the proboscis conditioning technique.

The principal disadvantage of this technique is that the contraction response resembles both the conditioned response (CR) and UR. As I will discuss in chapter 8, the conditioning of alpha responses presents some serious problems of interpretation. The behavioral researcher can

Figure 3-25

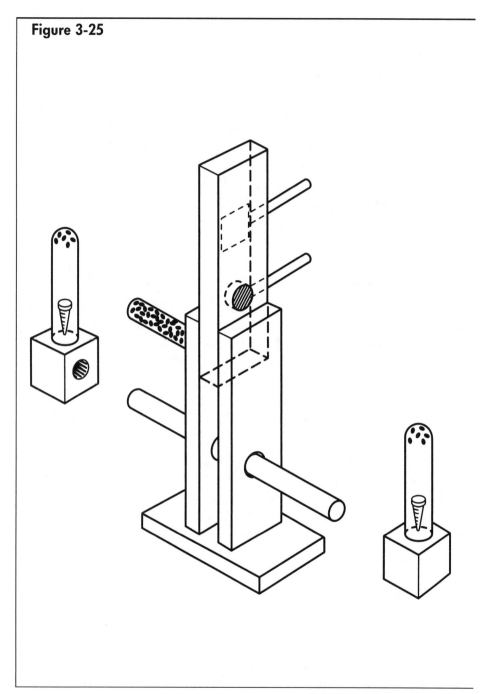

An olfactory conditioning device for flies (overview). From "Classical Conditioning and Retention in Normal and Mutant *Drosophila*" by T. Tully and W. G. Quinn, 1985, *Journal of Comparative Physiology (A)*, *96*, p. 157. Copyright 1985 by Springer-Verlag. Adapted with permission.

Figure 3-26

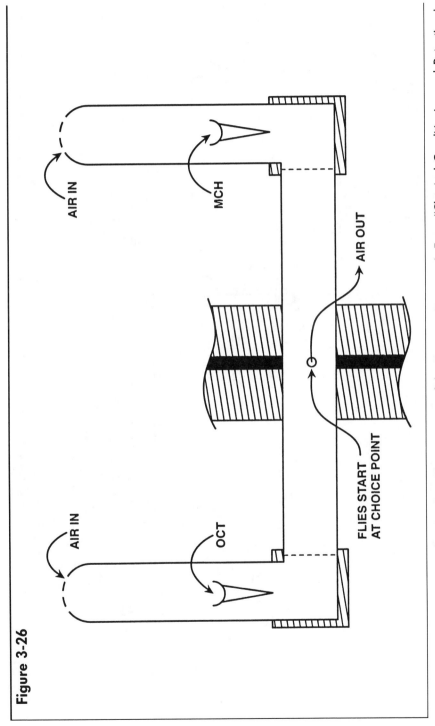

An olfactory conditioning device for flies (close-up view of the T-maze component). From "Classical Conditioning and Retention in Normal and Mutant *Drosophila*" by T. Tully and W. G. Quinn, 1985, *Journal of Comparative Physiology (A), 96*, p. 157. Copyright 1985 by Springer-Verlag. Adapted with permission.

Figure 3-27

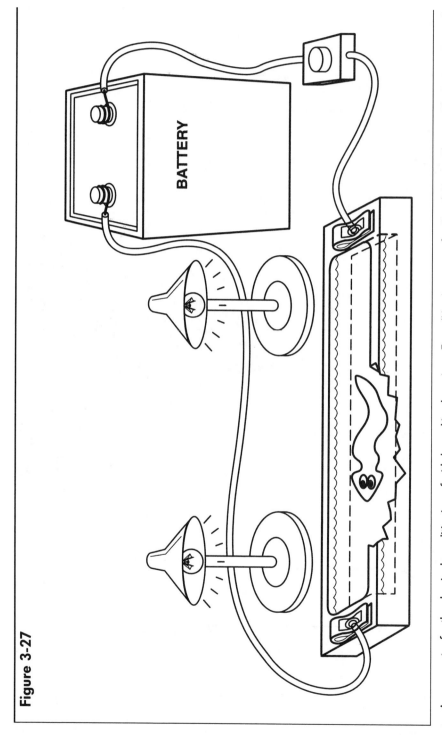

BATTERY

An apparatus for the classical conditioning of withdrawal in planarians. From "An Apparatus for Conditioning Planaria" by J. V. McConnell, P. R. Cornwell, and M. Clay, 1960, *American Journal of Psychology, 73*, p. 619. Copyright 1960 by the University of Illinois Press. Adapted with permission.

Figure 3-28

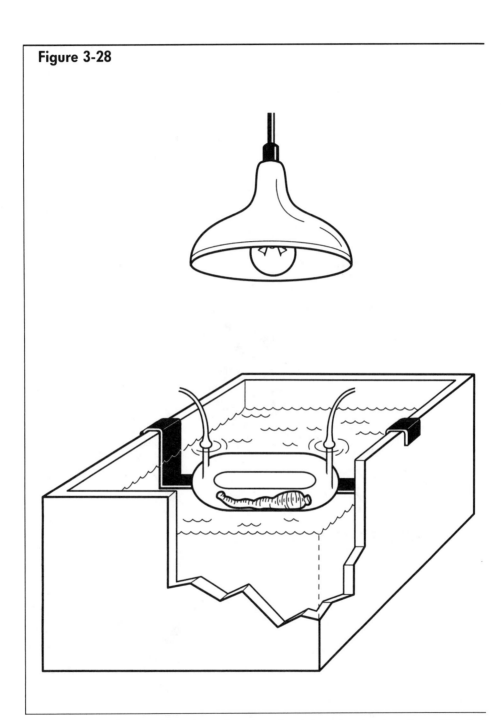

An apparatus for the classical conditioning of withdrawal in leeches. From "Classical Conditioning in the Leech *Macrobdella ditetra* as a Function of CS and UCS Intensity" by T. B. Henderson and P. N. Strong, Jr., 1972, *Conditional Reflex*, *7*, p. 211. Copyright 1972 by Transaction Publishers. Adapted with permission.

Figure 3-29

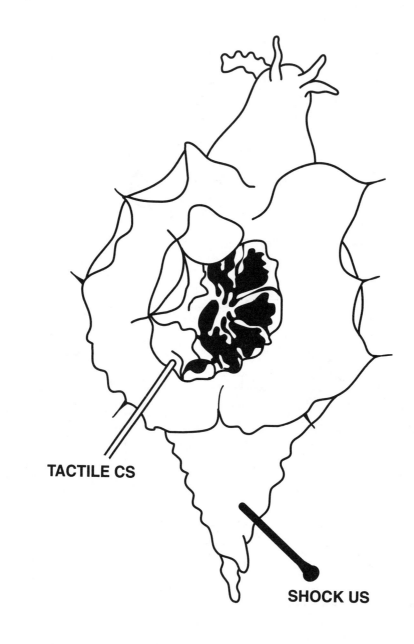

TACTILE CS

SHOCK US

An apparatus for the classical conditioning of withdrawal in *Aplysia*. From "Classical Conditioning in a Simple Withdrawal Reflex in *Aplysia californica*" by T. J. Carew, E. T. Walters, and E. R. Kandel, 1981, *Journal of Neuroscience, 1*, p. 1428. Copyright 1981 by the Society of Neuroscience. Adapted with permission.

easily interpret the pairing-specific increase in the amplitude of a withdrawal response to instrumental conditioning: The CS elicits a withdrawal response that is reinforced by the US. The reward in this situation may come from the ability of a contraction to lessen the effect of aversive stimulation. It was precisely this argument that largely discredited the planarian classical conditioning work. In addition, the technique has so far been restricted to aversive situations in which the US is shock, air puff, or intense light.

Withdrawal techniques are available for planarians (Thompson & McConnell, 1955), earthworms (Ratner & Miller, 1959a, 1959b), leeches (Henderson & Strong, 1972), *Aplysia* (Carew, Walters, & Kandel, 1981), and *Hermissenda* (Alkon, 1975). The withdrawal technique can also be used to train individual appendages such as the eye withdrawal reflex in the crab (Abramson & Feinman, 1988). Figure 3-27 depicts a withdrawal conditioning situation for planarians (Thompson & McConnell, 1955), Figure 3-28 for leeches (Henderson & Strong, 1972), and Figure 3-29 for *Aplysia* (Carew et al., 1981).

Summary

In this chapter, I have tried to impress upon you the importance of an apparatus in the design and analysis of an experiment. In addition, the characteristics of a good apparatus were discussed, and tips were provided on how to use a variety of devices commonly found in the invertebrate literature.

Discussion Questions

- What are the characteristics of a good apparatus?
- What factors should be considered in designing an apparatus?
- How are an activity wheel and an actograph used in studies of instrumental learning?
- What are the advantages and disadvantages of the maze technique?
- What are the advantages and disadvantages of the lever-press situation?
- What modifications are necessary to adapt an apparatus so that it is useful for neurophysiological investigations?

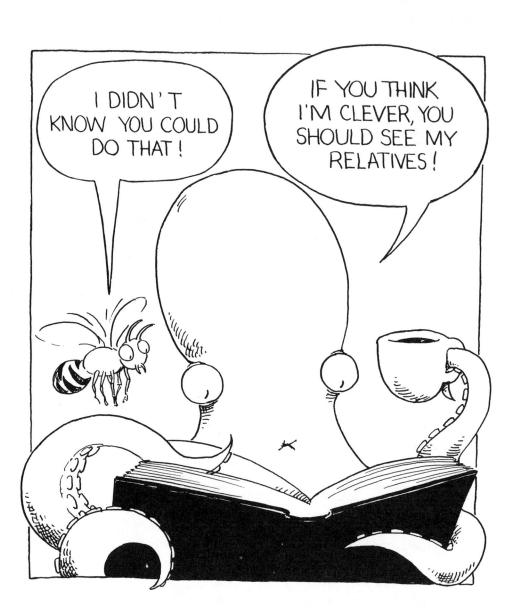

4 Habituation and Sensitization

Preview Questions

- What are some general considerations in the study of habituation and sensitization?
- What are habituation and sensitization?
- Why study habituation and sensitization?
- What controls should be employed in a study of habituation and sensitization?
- What are the essential elements in a study of habituation and sensitization?
- What stimuli have been used in invertebrate experiments on habituation and sensitization?
- What experimental variables have been manipulated in invertebrate studies of habituation and sensitization?

The motivations for the study of invertebrate learning are many and varied. We may, for example, be excited at the possibility of using invertebrates to evaluate a theoretical principle, trace the development of some phenomena through the animal kingdom, test a specific hypothesis, explore a functional relationship, or simply satisfy a curiosity. In previous chapters, I have discussed some of the techniques involved in the study of invertebrate learning and how the study of invertebrate learning is a cooperative enterprise spanning many scientific disciplines.

In this and the next three chapters, I will take a walk through the invertebrate learning literature and discuss some major findings and point out some of the research areas that can benefit from additional experimentation. The vertebrate learning literature will serve as a guide to identifying weaknesses and strengths in the invertebrate learning litera-ture. Using the vertebrate literature as a yardstick to measure the effec-tiveness of the invertebrate literature is more than a matter of conven-ience. As we saw in the last chapter, the powerful techniques developed to study learning of vertebrates are often extended to the invertebrate level. The classification of invertebrate learning into habituation, sensi-tization, and classical, instrumental, and operant conditioning is borrowed directly from terminology used to describe vertebrate performance. More-over, the rationale for many who study the learning of invertebrates is that the results of such experiments will reveal something about the way vertebrates, including humans, learn.

This chapter will examine nonassociative learning, specifically, ha-bituation and sensitization. Before proceeding, it is important to recall from chapter 1 that learning—whether it is associative or nonassocia-tive—is a hypothetical concept. Learning is never observed, it is inferred from changes in behavior. Bitterman (1967b) has described learning as "our name for a process (or set of processes) assumed to be responsible for certain of the observed changes. The nature of the process must be inferred from detailed studies of those changes and the variables to which they are related" (p. 139). This chapter will also discuss some of the necessary control procedures and focus on where—in my view—addi-tional research effort can be directed at the behavioral level. An overview of the cellular mechanisms of nonassociative learning will be presented in chapter 7.

What Are Habituation and Sensitization?

In the struggle to cope with the challenges of life, all animals—vertebrate and invertebrate alike—must be able to adapt to a constantly changing environment. Two of the oldest behavioral mechanisms developed to meet this challenge are known as *habituation* and *sensitization*. The exper-imental operations for producing habituation and sensitization are iden-tical. In each case, a stimulus is presented repeatedly. If some property of the response decreases, habituation has occurred. If the response in-creases, sensitization has occurred. Examples of these phenomena in our everyday lives are easy to find. Those of us who live in urban environ-ments, for example, are thankful that the habituation process allows us to tolerate, if not withstand, the constant barrage of unwanted noise emanating from an assortment of jackhammers, early morning garbage collection, traffic, and car alarms. Unfortunately, the repeated presentation of these same stimuli can also lead to sensitization—that is, we become annoyed.

Habituation refers to the decrease in amplitude, probability, or a change in the topography of a response to a monotonously repeated stimulus. Stimuli that no longer transmit any significant information, such as the presence of a predator or the location of a food source, tend, over time, to be ignored. The reduction in response strength to stimuli that initially elicited a host of reactions is the fundamental characteristic of habituation. Arguably, habituation is the most thoroughly studied example of nonassociative learning. In his bibliography of habituation experiments, for example, Leibrecht (1972, 1974) lists 1,684 experiments.

Sensitization is, in essence, the opposite of habituation and refers to an increase in frequency or probability of a response and is often accompanied by a decrease in latency and a lower threshold following the monotonous presentation of a usually strong stimulus. In addition to counteracting the influence of habituation, sensitization is important because it gives the animal an ability to increase the frequency of innate reactions. Increasing the frequency of innate behavior as a function of experience is a preliminary form of associative learning.

Just as there are various types of associative learning, there are various types of habituation and sensitization. To date, at least two kinds of habituation are generally recognized: short-term and long-term. The principal difference between the two is the length of retention. Short-term habituation refers to the response decrement within a testing session and may last from a few seconds to minutes to hours. In contrast, long-term habituation refers to the response decrement between training sessions and may last from hours to days to months.

Sensitization can also be divided on the basis of short-term and long-term retention and, in addition, can be subdivided further into nonassociative sensitization and conditioned sensitization (Razran, 1971). Nonassociative sensitization involves a single, usually intense stimulus such as an electric shock or vibration that elicits an innate response. An everyday example of nonassociative sensitization is the startle reactions evoked by the "slice and dice" horror films so popular today. Nonassociative sensitization is often used as a control in habituation experiments to ensure that the reduction in the response one is measuring is not the result of sensory adaptation or motor fatigue. When sensitization is used in this way, it is known more commonly as *dishabituation*.

The second category of sensitization is known as *conditioned sensiti-*

zation. This is an increase in the probability of an innate reaction based on the pairing of two stimuli. This type of sensitization is commonly known as *alpha conditioning*. There is some debate in the literature regarding the relationship between alpha and classical conditioning (Schreurs, 1989). For a review of the various types of nonassociative and associative sensitization, consult Razran (1971).

In contrast to most other types of behavior modification, the definitions of habituation and sensitization are remarkably consistent from experimenter to experimenter. If, for example, a group of psychologists, zoologists, and physiologists were asked to define habituation, they would probably provide a definition similar to that offered by J. D. Harris (1943), in which habituation is defined as a "response decrement as a result of repeated stimulation" (p. 385). The same is true of sensitization, which Teyler (1984) has defined as "an augmentation of a response to a stimulus" (p. 176). What is also remarkable about habituation and sensitization is that they are ubiquitous throughout the animal kingdom. They even appear, for instance, in experiments in which the "animals" consist of single cells or isolated ganglia. In terms of the evolution of learning mechanisms, habituation and sensitization may well be the most basic process for behavior modification.

Why Study Habituation and Sensitization in Invertebrates?

The study of habituation and sensitization is interesting for a number of reasons. First, habituation and sensitization experiments are easy to perform—whether the animal is freely moving or restrained, semi-intact, or "missing"—save for an isolated portion of a nervous system. Second, habituation and sensitization share many properties with more complex learning phenomena, such as the ability of the response to recover over time; creating new behavior patterns; improvement in performance over successive sessions; and sensitivity to such training parameters as intensity, frequency, and pattern of stimulation. Third, there are several well-defined characteristics (to be discussed later) that can be compared across species, such as the waning of the withdrawal response to light in planarian, earthworm, and mollusc. The characteristics of habituation and sensitization can be compared not only across species but also across the successive stages of a research preparation, as when we study sensitization in the intact *Aplysia*, a semi-intact specimen, and progress to isolated portions of the *Aplysia* nervous system. It has probably occurred to you that results obtained at each stage of your preparation can also be compared with analogous stages from other research preparations, such as the performance of semi-intact preparations of crayfish and *Aplysia*.

The significance of habituation and sensitization should not be underestimated. Though not as glamorous as the behavior change associated with classical or operant conditioning, their behavioral manifestations are just as adaptive. In addition, it must be kept in mind that for many invertebrates, this represents the only type of behavior modification. Habituation and sensitization increase the chances of survival and reproductive ability by minimizing wasted energy and by reducing the occurrence of maladaptive behavior. Wyers, Peeke, and Hertz (1973) and Papaj and Prokopy (1989) have identified the following adaptive features of habituation:

1. Alarm and escape reactions are restricted to the infrequently occurring stimulus patterns that characterize a predator's approach.
2. Orientation, exploratory, and defensive behavior that may compete with behaviors that enhance reproduction are reduced in familiar environments.
3. Innate and acquired behavior becomes restricted to appropriate situations.
4. Consummatory and other forms of behavior become possible in unpalatable but otherwise suitable situations.

Sensitization is also not without advantages. Wells (1967), Razran (1971), and Dyal (1973) have identified the following adaptive features of sensitization:

1. A mechanism is provided for reversing the effects of habituation.
2. The organism remains active and responsive to salient stimuli in hostile or novel environments.
3. Adaptive behaviors such as predator avoidance and food seeking emerge without the need for the extensive training characteristic of associative learning.

What Controls Should Be Used?

Before it can be concluded that a reduction in response can be attributed to habituation, four possible complications must be addressed: (a) base rate of responding, (b) sensory adaptation, (c) fatigue, and (d) temporal conditioning or sensitization (Ratner, 1970). In addition to the four sources identified by Ratner, I would add (e) general experience with the training task and (f) presence of pheromones. Although these controls are designed for habituation experiments, they are also applicable to studies of sensitization.

Base Rate of Responding

Be aware that many of the responses that habituate occur without any noticeable stimulation. This is common especially in invertebrate experiments where the response that is habituated involves some type of movement (such as eye withdrawal in the crab) and motion (in earthworms, planarians, and hydras). Before you can conduct, and accurately interpret the results of, any habituation experiment, it is important to know the rate, duration, and temporal pattern of the response that is to be habituated or sensitized. It is also important to determine if the change in behavior is the result of maturation or development. Given the rapid progress of invertebrate life cycles, such a possibility must be taken seriously. To establish a base rate of responding, add a control group to your experimental design that is placed in the training situation but not given any habituation training. Record the data as you would for a training run.

This type of control is analogous to establishing the "operant rate" of a response in operant conditioning experiments. In other words, you want to determine the frequency of response in the absence of any experimental manipulation. Such a control or reference group is also useful to detect the presence of rhythmic fluctuations in behavior due to, for example, seasonal variations and circadian rhythms. The detection of periodicities in invertebrate behavior is critically important if habituation and sensitization are to be properly assessed. Not to do so raises the possibility that the decrease (or increase) in responsiveness might have coincided with a behavioral cycle that affects responsiveness.

Sensory Adaptation

Another source of confusion in habituation experiments is the decrease in responsiveness in sensory organs subjected to intense or prolonged periods of stimulation. Two major procedures have been used to rule out the influence of sensory adaptation. First, you can select an intertrial interval—the time between presentations of the habituation stimulus— that is long enough to allow the effect of adaptation to wear off. If long intertrial intervals are not practical, a test trial procedure can be substituted in which habituation is assessed not during training but during test trials administered sometime after training. Select a time interval between training and testing that is long enough for adaptation to dissipate.

Effector Fatigue

A response progressively decreasing over the course of training may represent habituation, but alternatively, it may indicate that the effector mechanism(s) responsible for the expression of the response is not able to function. To separate the effects of fatigue from habituation, it is common practice to give the animal a test trial(s) using a second stimulus that elicits the same response. If there is a response to this other stimulus, the effect of fatigue may be ruled out. Gardner (1968), for example, was able to demonstrate that his worms were still able to respond by squeezing them with tweezers. Dishabituation is probably the most widely used control to assess the influence of fatigue in habituation experiments.

It should be mentioned that in most invertebrate studies of habituation, the stimuli used for dishabituation are often quite strong, such as electric shock, intense vibration, or a sharp pinch. Keep in mind that there is no reason to assume that such radical and intense stimuli do not also influence effector mechanisms that are fatigued. This is not often recognized in studies of invertebrate habituation. It does not take much imagination, for instance, to envision a scenario in which an earthworm that is too fatigued to respond to vibration will certainly do so following an intense electric shock.

A better method of control—in my opinion—is to use the same control procedures employed for sensory adaptation, that is, long intertrial intervals or test trials. If this is not possible, lower the intensity of your original stimulus, or in order to reduce the effects of stimulus generalization, use a second stimulus that is from a different sensory modality. In regard to the first method, it is interesting to note that a decrease in the intensity of stimulation has never been effective in producing dishabituation in invertebrates. A planarian, for example, that has ceased responding to light can be made to do so following an intense shake but not by reducing the intensity of the light during dishabituation trials. Why this is so is an interesting problem worthy of investigation.

Temporal Conditioning

When stimuli are presented on a regular basis as is often done in habituation experiments, the possibility exists that the animal may make the required response (or one in the opposite direction—sensitization) in anticipation of the stimulus. Such anticipation is common in vertebrate studies of temporal conditioning (a form of classical conditioning) but has not been demonstrated using invertebrates. Temporal conditioning would retard the rate of habituation, reduce the amount of habituation, and confound retention functions. Such anticipation is readily assessed by employing a control group that uses a variable intertrial interval; or instead of using a control group, be on the lookout for any responses occurring near the time of stimulus presentation.

General Experience in the Training Situation

In invertebrate studies of habituation and sensitization, the most common measure of retention is the savings or relearning score, in which performance is compared between training sessions. To take a simplified example, a planarian is found to contract to a pulse of light 70 out of the first 100 training trials. The next day, a second training session reveals that the animal responded on only 20 of the 100 trials. The 50 trials on which the animal failed to respond is considered a measure of retention. Such a savings score would indicate that the animal has retained information specific to the task. But has it?

At first glance, such performance appears to represent an impressive feat of memory. Consider, however, that the improvement from one training session to the next may be overestimated because the first session reflects influences of sensitization that are associated with unfamiliarity with the test situation, not present during subsequent sessions. To eliminate this possibility, a control group is added in which animals are placed in the testing situation for the same amount of time as animals in the experimental group, except that control animals receive no training. In the given example, the control animals would be placed in the apparatus for a period of time equalling the time needed to run 100 trials. Habituation training for these control animals would begin in the second session—24 hours after being exposed to the apparatus. The rate of habituation in the control animals during this session should approximate that found in the experimental animals—approximately 70 out of 100 trials. If, on the other hand, you find that the control animals contracted on 20 of the 100 trials—that is, resembling the second session performance of the experimental animals—even though the control animals received only one session of training, then the evidence for retention is difficult to support.

Pheromones

It is important to be aware that invertebrates emit various types of chemical signals and that these signals may influence habituation. Earthworms, for example, emit an alarm pheromone when stressed, and it is not unreasonable to assume that such pheromones are released into the experimental apparatus when the animal is placed inside and that the presence of such pheromones influences the rate of habituation. Response decrements to pheromones and other motivational agents, such as changes in hormone levels and hunger, are not considered examples of habituation. One way to gauge the influence of pheromones (or other chemical signals) is to place a new test subject in the same apparatus just vacated by the experimental subject. Assessing the role of pheromones and other types of motivational variables in habituation research with invertebrates is another topic that needs investigation.

What Are the Essential Elements in a Study of Habituation and Sensitization?

In this section, the components of a habituation and sensitization experiment are listed. (These components are also applicable to studies of associative learning in invertebrates.) The goal of this section is to present a guide that you can use to organize and characterize habituation and sensitization experiments. It is also useful as a guide to direct future research. The components selected have been shown to affect the findings of nonassociative learning in vertebrates. Selected examples of how these elements influence nonassociative learning in invertebrates are cited in parentheses. Detailed information is available by consulting the review articles cited in Table A-1.

What Stimuli Have Been Used in Invertebrate Experiments of Habituation and Sensitization?

Table 4-1 presents a partial, yet representative list of some invertebrate research preparations used in the study of habituation and sensitization. Additional references can be found in Table A-1.

Table 4-1

Habituation and Sensitization Stimuli and Responses

Animal	Stimulus	Response	Reference
Protozoa	vibration	contraction	Wood, 1970a
Coelenterta	rotation	contraction	Rushforth, 1967
Nematoda	shock	contraction	Haralson & Haralson, 1988
	vibration	reversal of movement	Rankin & Broster, 1992
Platyhel- minthes	light	contraction	Westerman, 1963a, 1963b
	vibration	contraction	Applewhite & Moro- witz, 1967
Annelida	vibration	contraction	Gardner, 1968
	light	contraction	Evans, 1969a
	mechanical shock	contraction	Evans, 1969b
	light	contraction	Lockery, Rawlins, & Gray, 1985
	tactile	swimming	Debski & Friesen, 1985
	tactile	contraction	Stoller & Sahley, 1985
Chelicerata	motion	eye movement	Land, 1971
	air puff	unit recording	Lahue & Corning, 1971
	vibration	movement	Szlep, 1964
Crustacea	tactile	tail flip	Krasne, 1969
	scent	chemosensory	Daniel & Derby, 1988
	shadow	escape	Brunner & Maldonado, 1988
	tactile	claw opening	Hawkins & Bruner, 1981
	tactile	eye movement	Appleton & Wilkens, 1990
Insecta	ultrasound	startle	May & Hoy, 1991
	motion	cardiac	Thon & Pauzie, 1984
	air puff	escape	Baxter, 1957
	poison	feeding	Szentesi & Bernays, 1984
Mollusca	sucrose	feeding	Braun & Bicker, 1992
	tactile	tentacle move- ment	Bruner & Tauc, 1965
	tactile	gill movement	Peretz, 1970
	tactile	siphon move- ment	Carew et al., 1972

What Experimental Variables Have Been Manipulated?

Thompson and Spencer (1966) identified nine characteristics of habituation based exclusively on vertebrate data. I will use these characteristics to demonstrate some of the parametric manipulations that have been performed with invertebrates. Although these characteristics are not universally accepted (see Graham, 1973; Hinde, 1970; Ratner, 1970), they are useful in facilitating quantitative comparisons across species and across research preparations. Hinde (1970), for example, suggests that another additional characteristic be added to recognize that habituation proceeds more rapidly when the time interval between presentation of the stimuli is short rather than long. Ratner (1970) agrees that the list be expanded to include the finding that consummatory behavior habituates more slowly than orienting behavior. For examples, see item 7 under Response measures (see Exhibit 4-1). The additional criteria proposed by Hinde and Ratner have seldom been examined using invertebrates and remain a topic for investigation.

The criteria of habituation proposed by Thompson and Spencer (1966) are also applicable to studies of sensitization and nicely illustrate the value of a polythetic approach to behavior analysis. The polythetic approach attempts to define a particular behavioral category in terms of multiple characteristics. For example, before you can conclude that a response decrement observed to vibration in earthworms and planarians is an example of habituation, you need to conduct additional comparisons. Such comparisons might involve determining whether the two animals show spontaneous recovery, stimulus generalization, intensity effects, or any of the remaining five criteria outlined by Thompson and Spencer (1966).

Before moving on to some of the data, I should mention that invertebrate studies of habituation typically examine no more than a few of the criteria proposed by Thompson and Spencer. The purpose of many invertebrate habituation experiments often appears to be simply a demonstration with no systematic analysis of the variables known to influence habituation. Some notable exceptions are the works of Carew et al. (1972) on *Aplysia*, Brunner and Maldonando (1988) on crabs, and May and Hoy's (1991) work on crickets. These experiments represent excellent examples of how to conduct a habituation experiment.

The lack of a consistent application of the polythetic approach to invertebrate studies of habituation has, with the exception of simple system work, failed to produce a coherent picture, although many isolated

Exhibit 4-1

Essential Elements in the Study of Habituation and Sensitization

- Subject variables
    1. Species comparison (Evans, 1969a, 1969b)
    2. Sex (Walker, 1972)
    3. Age (LeBourg, 1983; Peretz & Lukowiak, 1975; Zolman & Peretz, 1987)
    4. Developmental stages (Rankin & Carew, 1988)
    5. Intact, free-behaving animal (Carew, Pinsker, & Kandel, 1972)
    6. Semi-intact or isolated preparations (Peretz, 1970)
    7. Type of housing: isolated or group (not investigated)
    8. Prior experience (Clark, 1960b)
    9. Sensory capabilities (not systematically manipulated)
    10. Neuroanatomical organization (Eisenstein, Brunder, & Blair, 1982)
- Environmental variables
    1. Characteristics of apparatus (not systematically manipulated)
    2. Naturalistic vs. laboratory environments (Carew & Kupfermann, 1974; Walker, 1972)
    3. Temperature (Gardner & Applewhite, 1970)
    4. Seasonal variability (Lozada, Romano, & Maldonado, 1988).
    5. Ecological manipulations (Applewhite & Gardner, 1971)
- Response measures (intact, semi-intact, and isolated preparations)
    1. Frequency of response (Applewhite & Morowitz, 1967)
    2. Amplitude of response (Brunner & Maldonado, 1988; Rankin & Broster, 1992)
    3. Latency of response (Cook, 1971)
    4. Duration of response (Carew et al., 1972; Daniel & Derby, 1988)
    5. Number of responses (Kuenzer, 1958)
    6. Rate of habituation (Gardner, 1968)
        a. time or number of trials required to reach stability
        b. absolute amount of response decrement measured either within a given time period or by the number of trials
        c. relative amount of response decrement measured either within a given time period or by the number of trials
    7. Changes in topography (Balderrama & Maldonado, 1971; Gardner, 1968; Ratner, 1972)
    8. Individual differences (Clark, 1960b; Gardner, 1968; Rakitin, Tomsic, & Maldonado, 1991)
- Stimulus variables
    1. Single stimuli (most invertebrate studies)
    2. Multiple stimuli (Clark, 1960a; Osborn, Blair, Thomas, & Eisenstein, 1973).
    3. Concurrent stimuli (Rakitin et al., 1991; VanDeventer & Ratner, 1964)
- Controls
    1. Base rate of responding (Ratner & Gardner, 1968)
    2. Sensory adaptation (most invertebrate studies)
    3. Effector fatigue (most invertebrate studies)
    4. Temporal conditioning (no invertebrate experiments)
    5. General experience (Carew et al., 1972)
    6. Pheromones (Ratner & Boice, 1971)

facts have emerged. Our knowledge of the habituation process would be considerably advanced by applying the criteria of Thompson and Spencer—along with the additions proposed by Hinde and Ratner—to the multitude of invertebrate habituation experiments that by and large consist solely of demonstrations. The criteria proposed by Thompson and Spencer, along with several examples taken from invertebrate research, follow:

1. **"Given that a particular stimulus elicits a response, repeated applications of the stimulus result in decreased response (habituation). The decrease is usually a negative exponential function of the number of stimulus presentations"** (Thompson & Spencer, 1966, p. 18). This is a statement of the primary operational definition of habituation: a reduction in response strength to stimuli that initially elicited a reaction. Habituation has been demonstrated in all classes of invertebrates. Table 4-1 cites several examples.

   It is important to note that the vast majority of habituation experiments present group data. The individual data, however, may not show the negative exponential response decrement. In fact, many vertebrate habituation curves exhibit an initial period of increase responsiveness (Groves & Thompson, 1970). For an invertebrate example, see Rakitin, Tomsic, and Maldonado (1991), and Hawkins and Bruner (1981). It should also be noted that caution must be used in comparing habituation and sensitization curves across experiments because the shape of these curves critically depends not only on procedural differences associated with the 32 essential elements of habituation mentioned on p. 114 but also the nine criteria proposed by Thompson and Spencer.

2. **"If the stimulus is withheld, the response tends to recover over time (spontaneous recovery)"** (Thompson & Spencer, 1966, p. 18). Spontaneous recovery refers to the reappearance of a response that has decreased in strength, following the passage of time. Spontaneous recovery has been observed in all invertebrates. Spontaneous recovery has also been used as a measure of recall or memory. Retention of habituation has been demonstrated for up to 21 days in *Aplysia* (Carew, Pinsker, & Kandel, 1972), 1 day in the polychaete *Hesperonoë adventor* (Dyal & Hetherington, 1968), 4 days in the earthworm (Gardner, 1968), 6 hours in the protozoan (Wood, 1970a, 1970b), 4 days in the sea anemone

(Logan & Beck, 1978), 7 weeks in planarians (Westerman, 1963b), and 6 days in mantids (Balderrama & Maldonado, 1971).

3. **"If repeated series of habituation training and spontaneous recovery are given, habituation becomes successively more rapid"** (Thompson & Spencer, 1966, p. 18). This refers to the fact that extended training over a number of training sessions produces rapid and stronger habituation. This has been demonstrated in crickets (May & Hoy, 1991), leech (Debski & Friesen, 1985), and polychaetes (Dyal & Hetherington, 1968). A notable exception to this is the work of Appleton & Wilkens (1990), who found that some of their animals became sensitized with extended training sessions.

4. **"Other things being equal, the more rapid the frequency of stimulation, the more rapid and/or more pronounced is habituation"** (Thompson & Spencer, 1966, p. 18). This generalization has received some support in the invertebrate literature but needs additional research. One problem with this literature is that the majority of habituation experiments investigate only two interstimulus intervals. A wider range of intervals and species need to be investigated. A second problem is that most invertebrate studies assess the role of interstimulus interval by comparing habituation curves. The results of such studies are difficult to assess because they confound training and testing intervals. An earthworm that receives vibration every 5 seconds may differ from one that receives vibration every 10 seconds not because of the influence of the interstimulus interval but because the 5-second worm has twice the time to respond than the 10-second worm. As pointed out by Bitterman (1965) for studies of classical conditioning and by Davis (1970) for studies of habituation, all animals must be tested within the same interstimulus interval. An example of this control is found in the work of Thon (1987) and Rankin and Broster (1992).

With these two problems in mind, positive support for the influence of interstimulus interval can be found in work with many invertebrates, including protozoans (Rushforth, 1965, 1967; Wood, 1970a, 1970b), snails (Cook, 1971), crabs (Brunner & Maldonado, 1988), and polychaetes (Dyal & Hetherington, 1968). Partial support has been obtained in a study of *C. elegans* (Rankin & Broster, 1992), earthworms (Ratner, 1972; Ratner & Stein, 1965), and leeches (Boulis & Sahley, 1988). No support was obtained in a second earthworm study (Kuenzer, 1958) or in a study of the habituation of the optomotor response in blow flies (Thon, 1987).

5. **"The weaker the stimulus, the more rapid and more pronounced is habituation. Strong stimuli may yield no significant habituation"** (Thompson & Spencer, 1966, p. 19). Intensity effects have been reported for protozoans (Applewhite & Morowitz, 1966), Hydra (Rushforth, 1965), planarians (Brown, Dustman, & Beck, 1966), *Aplysia* (Pinsker, Kupfermann, Castellucci, & Kandel, 1970), leeches (Kaiser, 1954), crickets (May & Hoy, 1991), and crabs (Brunner & Maldonado, 1988; Lozada, Romano, & Maldonado, 1988). In general, these studies support the conclusion that the weaker the stimulus, the lower the initial level of the habituation curve and the fewer the trials needed for complete habituation. This conclusion, however, depends on the type of response investigated and the specifics of the experimental procedure. The results of intensity studies are also difficult to assess because many of the stimuli used in habituation experiments cannot be unambiguously scaled. In addition, as we discussed for the effects of interstimulus interval, most studies confound the training value of the stimulus with the test value. For the comparison of stimulus intensity to be more informative, it is necessary to test all animals with the same intensity of stimulation (Davis & Wagner, 1968).

6. **"The effects of habituation training may proceed beyond the zero or asymptotic response level"** (Thompson & Spencer, 1966, p. 19). This has seldom been investigated in invertebrates. Gardner (1968) appears to have performed the only experiment on this topic, and no effect was found in earthworms.

7. **"Habituation of response to a given stimulus exhibits stimulus generalization to other stimuli"** (Thompson & Spencer, 1966, p. 19). This characteristic has not been investigated thoroughly using invertebrates. Experiments demonstrate that habituation can generalize in, for example, some species of worms (Clark, 1960a; Kuenzer, 1958), snails (Cook, 1971), crustaceans (Brunner & Maldonado, 1988), insects (Braun & Bicker, 1992; May & Hoy, 1991; Thon & Pauzie, 1984), and protozoans (Wood, 1971, 1972, 1973). It should be mentioned that for generalization studies to be meaningful, differences in intensity between the original training stimulus and the test stimuli must be carefully equated.

8. **"Presentation of another (usually strong) stimulus results in**

recovery of the habituated response (dishabituation)" (Thompson & Spencer, 1966, p. 19). With the exception of work with aneural organisms such as *Stentor* (Eisenstein & Peretz, 1973; Wood, 1970a), dishabituation is in all invertebrate groups. As you may recall from the discussion of control procedures, dishabituation is often incorporated into the design of experiments to rule out the influence of other types of response decrements, such as sensory adaptation. In contrast to vertebrate studies of dishabituation in which stimulus change is effective, dishabituation has been produced in invertebrate experiments with only intense stimulation. The danger is that the increased responding may not be the result of novelty but rather a function of the intensity of the stimuli used to produce dishabituation.

9. **"Upon repeated application of the dishabituatory stimulus, the amount of dishabituation produced habituates (this might be called *habituation of dishabituation*)"** (Thompson & Spencer, 1966, p. 19). This, too, has seldom been investigated systematically using invertebrates. Habituation of dishabituation has been demonstrated in snails (Humphrey, 1930), insects (May & Hoy, 1991), and crustaceans (Brunner & Maldonado, 1988).

Summary

In this chapter, I have discussed the nonassociative learning paradigms of habituation and sensitization and identified the reasons that the study of these behavior modifications is important. In addition, I have discussed the importance of control procedures and identified the essential elements in the study of nonassociative learning.

Discussion Questions

- Compare and contrast habituation and sensitization.
- What is the importance of habituation and sensitization in the life of an invertebrate? Of a vertebrate?
- What control procedures are necessary in a study of habituation and sensitization?

- What type(s) of apparatus can be used in studies of habituation and sensitization?
- What are the essential elements in a study of habituation and sensitization?
- What experimental variables have been manipulated in studies of nonassociative learning?
- Using the Guidelines for Planning or Reporting Experimentation in chapter 2, design an invertebrate study on habituation.
- Using the Guidelines for Planning or Reporting Experimentation, go to the library, find an experiment on habituation, and analyze it according to the guidelines.

Classical Conditioning

Preview Questions

- What are some general considerations in the study of classical conditioning?
- What is classical conditioning?
- Why study classical conditioning?
- What controls should be employed in a study of classical conditioning?
- What are the essential elements in a study of classical conditioning?
- What stimuli have been used in invertebrate experiments on classical conditioning?
- What experimental variables have been manipulated in invertebrate experiments on classical conditioning?

To function successfully in a changing environment, animals must not only learn new behaviors but also call on reflexive responses in new contexts. Experimentally, this is demonstrated in the classical (Pavlovian) conditioning procedure. There are two general classes of classical conditioning experiment. In excitatory conditioning an US that evokes an UR or reflex is paired temporally with a novel stimulus, the CS. After continued CS–US pairings, the CS begins to evoke a CR. Classical conditioning is said to occur when responses to the CS occur with a higher probability, reduced latency, or greater magnitude over some baseline level, such as when the CS and US are unpaired, or alternatively, in situations in which two CSs are used—one of which is followed by the US. For example, to demonstrate classical conditioning in the blow fly,

Ricker, Brzorad, and Hirsch (1986) controlled for nonassociated effects by training the animal to discriminate between two different salt solutions (CSs), one of which was paired with sucrose (US).

In inhibitory classical conditioning, the CS inhibits the expression of the CR rather than elicits it. An inhibitory CS acquires associative strength by signaling the nonoccurrence of the US. Inhibitory conditioning is not measured directly as in the case of excitatory conditioning. Rather, inhibitory conditioning is assessed by the ability of the CS to inhibit a conditioned response that would have otherwise occurred. Studies of inhibitory classical conditioning are rare with invertebrates. Smith, Abramson, and Tobin (1991) were able to train honeybees to discriminate between two CSs, one associated with sucrose feeding and the other associated with a 10 V AC shock, if they responded to the sucrose US in the context of the second CS odor. Most subjects readily learned to respond to the odor followed by sucrose feeding and not to the odor associated with sucrose stimulation plus shock. Sucrose stimulation is normally a powerful releaser of the proboscis reflex, yet honeybees were able to inhibit this reflex in a context that would normally excite the reflex.

Classical conditioning procedures have been enthusiastically used in the study of invertebrate behavior since the early 1900s. As an example of one of these early demonstrations, Turner (1914) was able to train moths by pairing the sound of a tuning fork with electric shock. The impetus for many of these early studies was to assist in the creation of a "mental" phylogenic scale in the tradition established by Darwin and Romanes. Interest in invertebrate learning "waffled" during the 1930s and 1940s only to be taken up once again in a renewed effort to create a comparative analysis of learning. The decade of the 1950s witnessed the birth of such well-known Pavlovian paradigms as proboscis conditioning in insects and withdrawal conditioning in planarians and earthworms.

Since the 1960s, the interest in invertebrate studies of classical conditioning has grown steadily. Some of this interest is directed toward testing the limits of classical conditioning within the animal kingdom. Depending on one's definition of classical conditioning, these limits have been found to be quite broad, ranging from protozoans, flatworms, annelids, molluscs, and crustaceans to insects.

Another reason for the renewed interest in invertebrate studies is a greater appreciation of the type of behavior that can be conditioned. At one time, it was generally believed that classical conditioning was restricted to simple reflexes. Moreover, because it was thought that classical conditioning did not involve creating novel behavior, solving problems, or carrying out goal-directed activity, it was considered to be less important as a mechanism for behavior modification than instrumental and operant conditioning. This narrow view of classical conditioning has changed over the years. Experiments using vertebrates have expanded classical conditioning from the world of finger twitches, knee jerks, salivation, and eye blinks, to expectations, representations, and other concepts borrowed from the vocabulary of the cognitive psychologist. Today, it is generally believed that the animal learns that the CS predicts the onset of the US.

A third reason for the renewed interest in invertebrate classical conditioning concerns the quest for the cellular mechanisms of learning. The experimental rigor of the classical conditioning procedure combined with the accessibility of some invertebrate nervous systems is a powerful enticement for those interested in neuroscience.

What Is Classical Conditioning?

Classical conditioning is an example of associative learning in which the behavior of the animal is altered by the pairing of stimuli, one of which is effective in eliciting a biologically important reflex. In a broader sense, classical conditioning is a family of methods for the acquisition of associations between two or more stimuli or between stimuli and responses. A common feature of classical conditioning experiments is that the stimuli are presented independently of the subject's behavior. Classical conditioning is generally thought to represent the most basic of the associative learning mechanisms, one step above the nonassociative mechanisms of habituation and sensitization (see e.g., Hawkins & Kandel, 1984; Razran, 1971). Contemporary characterizations suggest, however, that the types of associations formed during classical conditioning are indistinguishable from those formed during instrumental and operant conditioning. As we saw in chapter 3, classical conditioning can be studied in a wide variety of apparatus such as running wheels and other types of actographs, modifying taste preferences, and in more confined situations such as proboscis conditioning.

Perhaps the most familiar example of classical conditioning is Pavlov's own research on conditioning salivary secretions in the now famous "Pavlovian dog." If you are not familiar with this work, imagine the taste of a freshly squeezed lemon as you slowly press the sour juice to your lips, or the excitement you experience at the sound of a smoke detector.

In both cases an association is formed between a nonsignificant or neutral stimulus (the spherical shape and color of the lemon and sound of a bell) and a biologically significant stimulus (citric acid and fire). When the neutral stimulus (the CS) reliably precedes the biologically significant stimulus (the US), we come to respond to the CS in a way that is different from the way we used to. In other words, we have established a CR. In our example, the innate reactions to citric acid and fire (salivation and fear) are the UR. Classical conditioning shares many properties with habituation and sensitization. These include extinction, spontaneous recovery, discrimination, generalization, and an increase in the strength of a behavior as a function of training.

In contrast to definitions of habituation and sensitization, the definition of classical conditioning is not always consistent from among experimenters. It is important to recognize this lack of consistency when evaluating invertebrate studies of classical conditioning. If, for example, we were to recall our group of psychologists, zoologists, and physiologists and ask them to define classical conditioning, we would most likely receive several different answers. Some psychologists such as Dickinson (1980) might stress that classical conditioning is the learning of relationships between cognitive events and that conditioning cannot be defined in terms of behavioral change. Other psychologists such as Gormezano (1984) stress that classical conditioning should be restricted to those procedures in which the CR is independent of the presentation of the US and that the CR be elicited by the same effector system that elicits the UR.

Within these extremes are definitions that stress that the CS must not elicit—prior to training—the response that is to be conditioned (Rachlin, 1970; Teyler, 1984), those that stress the contingency between the CS and US (Byrne, 1987), and still others that stress that classical conditioning is a procedure for creating a new reflex (Terrace, 1973). On the other hand, a zoologist or ethologist might consider classical conditioning to be the pairing of a "search image" with a sign stimulus or innate motor program (Gould, 1986). For the physiologist, the CS might be considered solely in terms of the electrical stimulation of afferent fibers. An excellent discussion on the problems in defining classical conditioning is available from Gormezano (1984).

Why Study Classical Conditioning in Invertebrates?

Perhaps the major reason for the study of invertebrate classical conditioning is its value as a comparative tool. The ease of conducting Pavlovian experiments on freely moving, restrained, semi-intact invertebrates, or on isolated portions of their nervous system, have stimulated their use in testing the generality of behavioral theories of classical conditioning and in developing neuronal models of the conditioning process. Much of the research on classical conditioning is conducted with the aim of discovering what changes occur in the nervous system during learning. As you will recall from chapter 2, this is known as the simple systems strategy. Classical conditioning procedures, however, have also seen increasing use as a bioassay to monitor the behavioral effects of pollutants (e.g., Mamood & Waller, 1990), and as a psychophysical tool to study the sensory capacities of insects (e.g., Getz & Smith, 1987; Towne & Kirchner, 1989). Little attention, however, is given to the comparative analysis of learning.

The lack of comparative data is detrimental not only to the psychology of learning but to the simple systems approach as well. In conducting simple systems research or evaluating the literature, do not assume that all classical conditioning procedures produce the same type of behavior modifications. A comparison of the parametric profiles (or polythetic profiles) of the various classical conditioning procedures has not been performed. Moreover, there is no agreement in the behavioral science community that these procedures tap the same behavioral process. Consider for a moment that the popular olfactory conditioning situation developed for fruit flies (Quinn, Harris, & Benzer, 1974), a variation of which we saw in chapter 3, has been considered over the years to represent classical conditioning (McGuire, 1984), avoidance (Dudai, 1977), and instrumental conditioning (Tully & Quinn, 1985)! The olfactory conditioning situation for flies is not the only example where the lack of behavioral details leads to confusion. Confusion also results when we fail to recognize that some procedures used to study classical conditioning may not be classical at all.

The problem of recognition and its impact for investigations of the underlying neural substrate has been considered by Gormezano, Kehoe, and Marshall (1983). As an example of the confusion that can result when we do not have a consistent idea of what classical conditioning is, let us take a moment and see how the scheme developed by Gormezano et al. (1983) compares with the available invertebrate classical conditioning procedures.

Gormezano et al. (1983) have divided the various vertebrate conditioning situations into four categories. The first category is known as the *CS–CR paradigm*. This paradigm, illustrated, for example, by the rabbit nictitating membrane situation and Pavlov's salivation experiments, represents the purest paradigm of classical conditioning. The hallmarks of these paradigms are that the CR appears in the same effector system as the UR; the experimenter has control of the US–UR complex.

The second category, known as the *conditioned stimulus–instrumental response (CS–IR) paradigm*, is represented by the conditioned suppression procedure and other transfer designs. In this case, there are CS–US pairings but no direct measurement of the CR. The effect of the pairing is measured by the ability of the CS to influence ongoing instrumental or operant behavior. For example, in one version of the widely used technique of conditioned suppression, an animal is placed in an apparatus where it receives several pairings of tone (CS) and shock (US). The effectiveness of the tone–shock pairing is not assessed by looking for a CR after each pairing but rather during a second experiment in which the animal has previously been taught to press a lever for food. During this lever-press experiment, the CS is now presented, and the extent to which it disrupts the pattern of lever pressing is assessed. As an everyday example of conditioned suppression, consider what happens to your performance when a demanding professor or some obnoxious authoritarian is standing over your shoulder. Rather than producing your normally flawless performance, you find yourself becoming inhibited.

The third category is known as *instrumental approach behavior* and is characterized by general activity to stimuli preceding food. This procedure is illustrated by general activity conditioning and instrumental runway and maze situations in which movement toward the food source is necessary. In such situations, measurement of the UR is confounded by the sight or smell of the food.

The final category is *autoshaping*. Autoshaping is a classical conditioning procedure in which response-independent stimuli, correlated with the availability of reinforcment, elicit approach behavior in the direction of the CS. This procedure is related to the CS–CR paradigm, with the exception that the CR is not from the same effector system as the UR.

Note that all of these four categories differ on how the CR is measured, the accuracy in which the CS and US are presented in a response-independent fashion, the nature of the target response, the amount of control the experimenter has over the training variables, and the degree to which the animal is restrained in the conditioning situation. The problem of comparing these four procedures can be solved only by a thorough behavioral analysis.

In Table 5-1, samples of the various invertebrate classical conditioning procedures discussed in previous chapters are cast in the scheme

Table 5-1

Categorization of Various Invertebrate Classical Conditioning Procedures

Paradigm	Animal	Reference
CS–CR	*Aplysia*	Lukowiak & Sahley (1981)
	Bee	Bitterman et al. (1983)
	Crab	Abramson & Feinman (1988)
	Crab	Mikhailoff (1922)
	Hermissenda	Lederhendler, Gart, & Alkon (1986)
	Housefly	Fukushi (1976)
CS–IR	*Drosophila*	Tully & Quinn (1985)
	Drosophila	DeJianne et al. (1985)
	Hermissenda	Crow & Alkon (1978)
Instrumental	Bee	Couvillon & Bitterman (1980)
Approach behavior	Bee	Sigurdson (1981a)
	Lobster	Fine-Levy, Girardot, Derby, & Daniel, (1988)
	Snail	Alexander et al. (1984)
	Paramecium	Gelber (1952)
Autoshaping	None	
Taste aversion	*Pleurobranchaea*	Mpitsos & Collins (1975)
	Limax	Sahley, Gelperin, & Rudy (1981)
Alpha conditioning	*Paramecia*	Hennessey, Rucker, & McDiarmid (1979)
	Planarian	Thompson & McConnell (1955)
	Earthworm	Ratner & Miller (1959a)
	Horseshoe crab	Smith & Baker (1960)
	Housefly	Fukushi (1979)
	Leech	Sahley & Ready (1988)
	Leech	Henderson & Strong (1972)
	Nereis	Inozemtsev (1990)
	Aplysia	Carew et al. (1981)
	Blow fly	Nelson (1971)

Note. CS = conditioned stimulus; CR = conditional response; IR = instrumental response.

developed by Gormezano et al. (1983). In addition, to these four categories I have added *alpha conditioning* and *taste aversion learning*. Alpha conditioning, you may recall from our previous discussion, is an example of conditioned sensitization and formally fits the definition of instrumental conditioning. In addition, there is no general agreement that taste aversion learning is an example of classical conditioning. Examination of this table reveals that only five procedures meet the criteria of the CS–CR paradigms. The other invertebrate procedures fall into the CS–IR paradigms, instrumental approach designs, taste aversion learning, and conditioned sensitization (alpha conditioning). There are no invertebrate examples of autoshaping.

The importance of using invertebrates to test the generality of classical conditioning and to create neuronal models of conditioning is firmly established. However, the participation of classical conditioning mechanisms in the daily life of many invertebrates is more difficult to assess. There are, of course, examples of the use of Pavlovian mechanisms in the control of invertebrate behavior, most notably in the social insects. Perhaps the most well-known is the learning of landmarks, orientation, and food sources in ants and bees. Much of the behavior critical to survival and reproduction in invertebrates, however, is under the control of innate behavior such as the kineses, taxes, and compass reactions of the sort proposed by Fraenkel and Gunn (1961). Moreover, nonassociative learning such as sensitization, habituation, and pseudoconditioning may play a greater role in modifying invertebrate behavior than do classical conditioning mechanisms (Wells, 1967). As we discussed in the section on habituation and sensitization, an advantage of nonassociative learning is that it does not need the extensive training characteristic of associative learning. Nonassociative learning, until recently, has not been as intensely researched as associative learning.

What Controls Should Be Used?

Before it can be concluded that the appearance of a CR is the result of the formation of an association between the CS and US, several alternative explanations must be eliminated. Chief among them are pseudoconditioning and sensitization. In addition to these we must also consider base rate of responding, temporal conditioning, and pheromones. The latter three groups are familar to you from our discussion of control groups in habituation and sensitization.

Pseudoconditioning

If a US is presented over the course of several trials and then a CS is introduced, a response resembling that elicited by the US will occur to the first presentation of the CS. This is not considered a CR because it was not the result of CS–US pairings. To estimate the amount of *pseudoconditioning*, include a control group that receives the same number of CSs and USs as the experimental group, but separated by an intertrial interval. It is customary when employing an unpaired control to keep the training period constant between the experimental and unpaired control groups. This is accomplished by using an intertrial interval that is half that used in the experimental group. For example, if the intertrial interval is one minute in the experimental group, it will be 30 seconds in the unpaired control group. Classical conditioning would be demonstrated if responses to the CS are greater in the group receiving paired CS–US presentations than in the group receiving unpaired presentations. A second way to estimate pseudoconditioning is to train an animal to discriminate between two CSs, one of which is paired with the US. Classical conditioning would be demonstrated if the animal is able to discriminate between them.

As I mentioned, pseudoconditioning should not be considered merely a control in classical conditioning experiments. The behavior modifications produced by pseudoconditioning are as important as those produced by classical conditioning and deserve additional attention by behavioral scientists (i.e., Evans, 1966; Kimble, 1961; Razran, 1971; Wells, 1967; Wickens & Wickens, 1942). How pseudoconditioned responses are formed is still a mystery. Such responses may reflect an internal sensitization process that causes a particular external stimulus to trigger a response; alternatively, such responses may be the result of a similarity between the US and the CS (Mackintosh, 1974).

Sensitization

All potential conditioned stimuli elict responses that may be potentiated by the presentation of the US. Sensitization is an increase in the conditioned response as a result of the presentation of the US. Interpreting the effectiveness of the CS–US pairing becomes especially difficult if one of the responses elicited by the CS serves as the index of learning. This is a common problem with invertebrate studies of classical conditioning and has been examined in some experiments by preceding conditioning trials with CS presentations. When responses to the CS no longer occur, CS–US pairings are begun. Preexposing the CS is an unacceptable procedure because one CS–US pairing is often sufficient to reinstate the response.

A second problem with exposing the invertebrate to CS presentations prior to CS–US pairings is that preexposure may produce latent inhibition that would influence the acquisition of classical conditioning. Sensitization can be assessed by employing an unpaired control group or by requiring the animal to discriminate between two CSs, one of which is paired with the US. Although inadequate to rule out sensitization and pseudoconditioning, employing CS-only and US-only control groups will provide an indication of whether repeated exposure to the CS potentiates the CR and whether repeated presentations of the US elicits responses resembling the CR.

Base Rate of Responding

As in habituation and sensitization, responses may occur during classical conditioning without any noticeable stimulation. Interpreting the success of training becomes problematic if the frequency of these spontaneous responses becomes so high as to occur during the presentation of the CS. To assess the influence of spontaneous responses, a blank or activity control group is employed in which responses are recorded during an imaginary CS. The imaginary CS is presented on the same schedule as that used in the experimental group. A second procedure is to use a within-subject design in which blank or activity trials are interpolated with CS–US trials.

Temporal Conditioning

In most invertebrate studies of classical conditioning, the interval between CS–US pairings is fixed. This is a bit surprising because the vertebrate data suggest that the passage of time can function as a CS. To eliminate the possibility that animals anticipate the CS, vertebrate experiments routinely vary the intertrial interval. If a variable intertrial interval cannot be incorporated into the experimental design, the amount of temporal conditioning can be estimated by employing a US-only control group in which the intertrial interval is the same as that used in the paired group. "CS" responses are measured during the time interval preceding the US.

Pheromones

The influence of pheromones in invertebrate classical conditioning experiments has not been systematically investigated. Their role in instrumental maze and runway experiments, however, is profound. In classical conditioning experiments, it is a standard procedure to clean the apparatus prior to the introduction of the subject. In the free-flying procedure, for example, the targets are washed following each visit of the honeybee. In proboscis and olfactory conditioning experiments, it is standard procedure to continuously ventilate the apparatus.

Calendar Variables

Seasonal variables and changes in local weather conditions have a strong influence on invertebrate behavior. These variables are difficult to manipulate, but they can be assessed by intermingling the running of the experimental and control groups. For example, if you have 20 experimental bees ready for proboscis conditioning and 20 control bees, do not run all 20 experimental bees before running the control animals. Rather, for every experimental animal, run a control animal.

Central Excitatory State

In conducting classical conditioning studies of proboscis conditioning in flies and bees, it is important to be aware of, and control for, what has been called the *central excitatory state* (CES; Dethier, Solomon, & Turner, 1965). A hungry fly or bee, when stimulated by sugar, will respond by immediately extending its proboscis. The problem you will face when interpreting the results of such a classical conditioning experiment is that the sugar used as the US will establish a temporary state of excitement in the animal's central nervous system, which will produce proboscis extension to stimuli (such as a CS) to which it would not normally respond. Such nonassociative effects will overestimate the level of conditioning.

Over the past decade, there has been considerable debate in the fly literature over the issue of CES and its role in the interpretation of learning in flies. An excellent discussion on this important topic is provided by Ricker, Hirsch, Holliday, and Vargo (1986). To control for CES, Ricker, Hirsch, et al. (1986) suggest several procedures. One is to use long intertrial intervals. The interval that you select must be long enough for the CES to dissipate. A second procedure is to incorporate into your experimental design a second stimulus that can be used to "discharge" a proboscis extension associated with the CES. This second stimulus is presented during the intertrial interval and before the presentation of the CS. In this way you can separate proboscis responses produced by the CES from those that are the result of the CS–US pairing. A third technique is to use discrimination learning in which the animal must distinguish between two CSs, one of which is paired with sugar.

Systematic Variation

When comparing the results of your classical conditioning experiments with the results in the literature, it is important to resist any tendency to generalize your findings to other species or to make the claim that you have uncovered a species difference. It is important to keep in mind that your procedures may differ in many significant ways from those used by other researchers and that the differences you detect may simply be the result of some procedural or organismic variable such as motivation level, sex, age, or training parameters. These and other variables must be systematically explored or varied before you can say with any confidence that you have uncovered a qualitative difference or similarity between your research and the research of others.

Experimenter Bias

When conducting experiments with invertebrates, it is important to note that there may be subtle differences among the researchers in your laboratory in the way that they present stimuli to the animal, handle the animal, and mount the animal in the conditioning apparatus. One way to control for experimenter bias is to run the experiments "blind," that is, the experimenter does not know the group he or she may be working on. Another method is to use several experimenters for the same group. Perhaps the best way is to automate your stimulus delivery and data recording systems. Given the lack of available commercial apparatus and standardized construction plans, this is not always possible.

What Are the Essential Elements in a Study of Classical Conditioning?

In this section, some of the key variables that influence invertebrate classical conditioning are briefly discussed. Detailed information on the elements of a classical conditioning experiment is available from any learning text such as Mackintosh (1974) and Mazur (1994).

Conditioned Stimulus

A CS is characterized by its inability to elicit the CR prior to CS–US pairings. In vertebrate experiments, a wide range of CSs are used. Some familar examples are tones, shapes, and colors. In contrast, the CSs used in invertebrate experiments are very limited and more often than not consist of touch, vibration, light, taste, and odor. In neuronal preparations, the "CS" is the electrical stimulation of afferent fibers. Many invertebrate researchers such as Roger and Galeano (1977) make a distinction between a CS used in behavioral experiments and those used in neuronal preparations. A sample of the type of CSs used for various invertebrates appears in Table 5-2.

Unfortunately, most of the CSs used in invertebrate behavioral experiments elicit the response that is to serve as the index of learning. One method that has been used in invertebrate studies to reduce these responses is to preexpose the CS prior to training. This method is not desirable for two reasons. First, although the response may appear to adapt, it often returns to its former strength following the first CS–US pairing. Second, the invertebrate may habituate to the CS, thereby reducing the effectiveness of the CS–US pairing.

An interesting question concerns the ineffectiveness of light as a CS in proboscis conditioning experiments and other situations in which an insect remains stationary, as, for example, when it is feeding (e.g., Walker, Baird, & Bitterman, 1989). Although light has been used as a CS in the classical conditioning of withdrawal responses in planarians and leeches, in each case the animal is moving within the apparatus, and the results of CS-only controls suggest that we are dealing with alpha conditioning. Light, however, has been effective in one *Aplysia* experiment and in one crab experiment in which alpha responses were not reported (Lukowiak & Sahley, 1981; Mikhailoff, 1922). Why light appears to be ineffective in insects and not in molluscs and crustaceans is not clear. Changes in magnetic field have also been ineffective in stationary invertebrates (Walker et al., 1989).

In designing invertebrate classical conditioning experiments, one of the biggest challenges you will face is selecting stimuli to serve as CSs. As a rule of thumb, CSs should not be too intense, must be sensed by the invertebrate, and should not elicit a response that resembles the CR.

Table 5-2

Samples of CSs and USs Used in Invertebrate Classical Conditioning Studies

Animal	CS	US	Reference
Aplysia	tactile	shock	Carew et al. (1981)
Aplysia	chemosensory	shock	Walters, Carew, & Kandel (1979)
Aplysia	light	tap	Lukowiak & Sahley (1981)
Bee	olfactory	food	Bitterman et al. (1983)
Crab	vibration	air puff	Abramson & Feinman (1988)
Crab	light	tap	Mikhailoff (1922)
Earthworm	vibration	light	Ratner & Miller (1959)
Fly	light	food	Fukushi (1976)
Fly	olfactory	food	Fukushi (1979)
Fly	olfactory	shock	Tully & Quinn (1985)
Fly	olfactory	food	Akahane & Amakawa (1983)
Fly	saline	food	Nelson (1971)
Hermissenda	light	rotation	Lederhendler et al. (1986)
Horseshoe crab	light	shock	Smith & Baker (1960)
Leech	tactile	shock	Sahley & Ready (1988)
Leech	light	shock	Henderson & Strong (1972)
Limax	taste	toxicosis	Gelperin (1975)
Lobster	chemosensory	predator	Fine-Levy et al. (1988)
Lobster	acoustic	shock	Offutt (1970)
Locust	olfactory	food	Simpson & White (1990)
Nereis	vibration	shock	Inozemtsev (1990)
Paramecium	vibration	shock	Hennessey et al. (1979)
Planarian	light	shock	Thompson & McConnell (1955)
Pleurobranchaea	taste	shock	Mpitsos & Davis (1973)
Sea anemone	light	shock	Haralson, Groff, & Haralson (1975)

Note. CS = conditioned stimulus; US = unconditioned stimulus.

Although these rules may seem obvious, it is often difficult to apply them at the invertebrate level. How do you determine, for example, whether an invertebrate is sensing your stimulus, or the stimulus is too intense? One method is to look for an orientation response.

In general, with the intensity of the US held constant, the strength of conditioning increases the more intense the CS. However, if the CS is too intense, conditioning will decline. CS intensity has not been systematically manipulated in many invertebrate experiments. In those cases where it has been manipulated, increases in CS intensity do increase the strength of conditioning in *Nereids* (Inozemtsev, 1990); *Paramecia* (Hennessey et al., 1979), and leeches (Henderson & Strong, 1972). Table 5-2 presents a representative list of some of the CSs and USs used in invertebrate studies.

An interesting variation of CS parameters occurs when two or more CSs are presented in a compound. The individual CSs forming the compound are known as *elements*. These elements can be presented successively or simultaneously. A popular simultaneous compound stimulus in free-flying honeybee experiments, to cite one example, is a single target consisting of both color and odor. If one element in a compound is more intense or salient than the other, most of the conditioning will be controlled by the more noticeable element. In the jargon of the conditioning experiment, the more intense element *overshadows* the less salient element. Salience is also detemined by the animal's prior experience with the individual elements. The overshadowing of one element as a result of experience with a second element is known as *blocking*. A third result that sometimes occurs with compound conditioning is that if the subject receives extended training with the compound stimulus, then neither component of the compound stimulus will, when tested in isolation, elicit a CR. Rather, a CR will be elicited by the compound but not by the individual elements that make up the compound. This phenomenon is known as *configural conditioning* (Mackintosh, 1974; Razran, 1971). Blocking, overshadowing, and configural conditioning have been extensively studied in free-flying honeybees (Couvillon & Bitterman, 1980, 1982), and blocking has been demonstrated in a taste aversion situation with *Limax* (Sahley, Rudy, & Gelperin, 1981).

Unconditioned Stimulus

A US reliably evokes a measurable response. The US–UR complex is known as the *unconditioned reflex*. The USs used in classical conditioning situations fall into two broad categories: *appetitive* and *aversive*. If the US is shock or other noxious stimulus, the classical conditioning situation is known as *classical defense* (or *aversive*) *conditioning*. When the US is food or some other rewarding event, the situation is known as *classical reward* (or *appetitive*) *conditioning*. Because the USs used in classical conditioning experiments cannot in principle be modified by the subject—CS–US pairings are presented independently of the animal's behavior—the determination of rewarding and aversive stimuli is made in instrumental or operant conditioning experiments. The majority of invertebrate studies of classical conditioning use aversive events like electric shock as USs. The notable exceptions are the proboscis conditioning situations in which appetitive stimuli such as sucrose are used.

When selecting a US and determining an effective intensity, keep in mind that it must reliably elicit a response throughout the course of training. For example, if the US is a weak electric shock, the animal will habituate and conditioning will be erratic. On the other hand, if food is used in an appetitive conditioning situation such as proboscis conditioning, care must be taken to ensure that the animal does not satiate to sucrose. It is also important to note that the US you select restricts the type of CR to those that are elicited by the US.

In general, with the intensity of the CS held constant, the strength of conditioning increases the more intense the US. The intensity of the US has not been systematically manipulated in many invertebrate experiments. In those cases where it has been manipulated, increases in US intensity does increase the strength of withdrawal conditioning in leeches (Henderson & Strong, 1972) and proboscis conditioning in honeybees (Bitterman et al., 1983).

An interesting variation of US parameters occurs when the animal is preexposed to the US prior to CS–US pairings. Vertebrate experiments indicate that such preexposure often retards conditioning. This US preexposure effect has been demonstrated in a free-flying experiment with honeybees (Abramson & Bitterman, 1986) and a taste aversion study with Limax (Sahley et al., 1981). Table 5-2 presents a representative list of some of the USs used in invertebrate studies.

Response Measurement

The acquisition of CRs has been assessed several ways. The most popular measurements involve recording the frequency, latency, amplitude, and probability of the CR. At the neuronal level, these measures have been applied to changes in electromyograms, membrane potentials, and action potentials. Most invertebrate studies employ only one response measure. A more complete picture of conditioning can be obtained by recording several measures simultaneously. The type of responses that will be recorded depends of course on the sensitivity of the available recording devices. Most of the early invertebrate conditioning experiments presented stimuli manually and used observational methods to record responses. Today, many of the invertebrate experiments have reached a level of sophistication that is equal to any vertebrate experiment.

When designing a classical conditioning experiment, a decision must be made whether to record CRs on every trial (known as the *anticipation method*), or only on some of the trials (known as the *test trial method*). There are advantages and disadvantages to both methods. To use the anticipation method, the CS must be long enough for the subject to respond before the onset of the US. An advantage of this technique is that the progress of conditioning can be tracked on a trial-by-trial basis. The latency of different response systems vary, however. This becomes a problem if the response that is to serve as the index of learning needs more time to express itself than the CS duration you select. For example, if it takes 1 second for the crab to withdraw its eye and the experimenter's CS duration is ½ second, it would not be possible to record CRs. To get around this problem, one could lengthen the CS to 3 seconds. However, in lengthening the CS, one risks using a CS–US interval that is ineffective in supporting classical conditioning.

A second solution to this problem is to use the test trial method. Here, conditioning is not measured on every trial; rather, during selected trials, the CS is presented alone and is of a longer duration than that used in training. The longer duration makes it possible to detect a CR. Returning to our crab example, one could maintain the 1-second duration and present unreinforced test trials, for instance, every five trials, in which the CS is 3 seconds long. The test trial method is preferred when short CS durations and short CS–US intervals are employed. A potential problem here, however, is that on test trials, the US is omitted and the CR is weakened through extinction.

You should also be aware that different response systems condition at different rates. Using the mean number of trials to the first conditioned response, Lennartz and Weinberger (1992) have discovered that there is a bimodal acquisition rate for several of the response systems used in vertebrate classical conditioning experiments. The first CR as measured by the galvanic skin response, condition suppression, blood pressure, respiration, pupil, and heart rate usually appears within the first 10 CS–US pairings. The appearance of the first CR in the nictitating membrane, limb/tail flexion, and eyelid conditioning procedure, however, do not usually appear until about 100 CS–US pairings.

Although the number of invertebrate experiments is small and it is substantially difficult to make comparisons because the conditioning parameters vary so widely among the invertebrates studied, a trimodal distribution is evident. The appearance of the first CR in taste aversion can occur within a single trial (Gelperin, 1975) and proboscis conditioning within five trials in the housefly and bee (Bitterman et al., 1983; Fukushi, 1979). Withdrawal conditioning, however, in *Aplysia* (Carew et al., 1981), earthworms (Herz et al., 1964), and the leech (Sahley & Ready, 1988) begins to appear within 10 trials. Flexion conditioning of eye movement in crabs appears after about 30 trials (Abramson & Feinman, 1988) and tail flexion in the horseshoe crab after about 80 trials (Smith & Baker, 1960).

Temporal Spacing of CS and US

When the CS and US have been selected, and you have determined how the conditioned response is to be measured, a decision must be made as to the temporal arrangements between these stimuli. The precise control provided by the classical conditioning situation over the onset and termination of the CS and US permits several well-known options. Each of these options affect the strength and type of conditioning. In general, the onset of the CS must precede the US if an increase in the CR is to occur.

In *delay conditioning*, the CS is presented prior to the US and continues until the onset of the US. In some examples of delayed conditioning, the CS overlaps the US, with the CS and US terminating together. The interval between the onset of the CS and the onset of the US is known as the *interstimulus interval* (ISI), or *CS–US interval*. Delay conditioning has been found to be the most effective procedure for increasing the efficacy of a CR. Both types of delay conditioning procedures have been used with invertebrates. However, the vast majority of invertebrate studies have been conducted with a CS that terminates with the US. The differences in delay conditioning procedures should be kept in mind when reviewing invertebrate studies of classical conditioning. For example, studies of withdrawal conditioning in *Paramecia* (Hennessey et al., 1979), planaria (Thompson & McConnell, 1955), the leech (Henderson & Strong, 1972), earthworm (Ratner & Miller, 1959a) and the proboscis conditioning experiments with various insects (i.e., Bitterman et al., 1983) have all used CSs that overlap the US. Studies using a CS that terminates immediately prior to the US are rare. An example can be found in the classical conditioning of the eye withdrawal reflex in the crab (Abramson & Feinman,

1988) and in a recent study of proboscis conditioning (Batson, Hoban, & Bitterman, 1992).

In *simultaneous conditioning*, the onset of the CS and of the US occur at the same time. At one time, vertebrate studies of simultaneous conditioning indicated that there is little or no association when CRs are measured on test trials in which the CS is occasionally presented alone. Recent work, however, suggests that associations may be formed (e.g., Rescorla, 1980; Rescorla & Cunningham, 1978). Simultaneous conditioning of invertebrates has been detected in the proboscis conditioning situation for bees (Batson et al., 1992) and houseflies (Fukushi, 1979) and in a withdrawal situation in *Aplysia* (Hawkins, Carew, & Kandel, 1986). Whether simultaneous pairing of the CS and US produces association in other invertebrate situations is not known.

In *backward conditioning*, the US *precedes* the CS. At one time, vertebrate studies revealed, as in the case of simultaneous conditioning, that little or no excitatory learning takes place. There is evidence, however, that backward conditioning may result in inhibitory conditioning. Studies of invertebrate classical conditioning have not determined whether backward conditioning results in inhibitory learning. *Inhibitory learning* refers to the supression of behavior that otherwise would have occurred. It is known, however, that backward conditioning does not produce excitatory conditioning in *Aplysia* (Hawkins et al., 1986), crabs (Abramson & Feinman, 1988), the leech (Sahley & Ready, 1988), planarians (Jacobson, Fried, & Horowitz, 1967), and flies (Fukushi, 1976; 1979).

In *trace conditioning*, the CS begins and ends before the US is presented. In general, if the interval between the CS and US is less then 5 seconds, it is referred to as *short-trace conditioning*. If the interval is longer then 5 seconds, it is referred to as *long-trace conditioning*. Trace conditioning is as effective as delay conditioning in producing CRs at short ISIs. However, as the ISI is increased, delay conditioning is more effective. There are few studies of trace conditioning in invertebrates. Hawkins et al. (1986) demonstrated in *Aplysia* that conditioning was most effective when the CS preceded the US by 0.5 seconds, was marginal at 1 second, and did not occur at 2, 5, or 10 seconds.

In *temporal conditioning*, the US is presented after a fixed interval of time has elapsed. There is no explicit CS. The "CS" is the passage of time. After extended training, the animal begins to respond near the onset of the US. The data on temporal conditioning in invertebrates are controversial. The only available evidence that invertebrates can time durations comes from the rather extensive literature in which free-ranging honeybees and ants are trained to come to a feeder at a particular time of day. (This literature has been extensively reviewed elsewhere by Gould, 1991; Schneirla, 1953; von Frisch, 1967.) It is not clear, for example, why temporal discrimination in bees is better in the morning and gets progressively worse as the day progresses (Moore & Rankin, 1983). In addition, failures to replicate have been noted in some of the early literature (Schneirla, 1953). Moreover, it is unclear how experimenters ensure they are not repeatedly counting the same animals and that the animals have not already sampled the feeding site before the experimenter has arrived to record data. Only one experiment has demonstrated temporal conditioning under laboratory conditions (Logunov, 1981). Logunov reported temporal conditioning to tactile stimulation of the pneumostome in the restrained snail. Unfortunately, the reliability of the data is difficult to assess because the report lacks some of the necessary methodological and statistical detail.

In contrast to the field experiments and the laboratory investigations of Logunov, temporal conditioning has not been demonstrated in the proboscis extension reflex or antennae movement in honeybees nor in the eye withdrawal reflex of the crab (Abramson, unpublished observations). In addition, temporal regularity has not been demonstrated in a fixed-interval free-operant experiment in crabs (Abramson, unpublished observations) nor in honeybees (Grossmann, 1973).

An interesting phenomenon associated with delay conditioning in vertebrates, known as *long-delay conditioning*, occurs when the CS–US interval is relatively long, such as an interval of several minutes. Early in training, CRs occur at the onset of the CS. However, as training continues, the CR begins to appear immediately prior to the US. This phenomenon is known as *inhibition of delay* and provides support for the view that vertebrates are able to discriminate time. Long-delay conditioning is an ideal technique to test the ability of invertebrates to form temporal discriminations. Unfortunately, such a procedure has not been attempted with invertebrates.

CS–US Interval

The CS–US interval is the period of time from the onset of the CS to the onset of the US and is also known as the interstimulus interval, or ISI (see p. 138). Interval selection is a most important variable in the classical conditioning experiment because it defines CS–US contiguity and therefore, the limits of classical conditioning. In general, the shorter the CS–US interval is, the better the conditioning. However, if the CS–US interval is too short, conditioning will be poor. Despite the importance of the CS–US interval, there are few invertebrate studies on this topic. Fukushi (1979) showed that the level of proboscis conditioning in houseflies decreased when the CS–US interval was 10 seconds rather than 5 seconds. It should be mentioned that the CS–US interval varies widely in invertebrate experiments. Intervals range from 30 seconds in *Hermissenda* (Lederhendler et al., 1986), 15 seconds in snails (Alexander et al., 1984), 5 seconds in the housefly (Fukushi, 1976), 10 seconds in the horseshoe crab (Smith & Baker, 1960), 6 seconds in the earthworm (Herz et al., 1964), 6 seconds in the bee (Bitterman et al., 1983), 5 seconds in Nereids (Inozemtsev, 1990), 1 second in the leech (Sahley & Ready, 1988), 1 second in the crab (Abramson & Feinman, 1988) and 0.5 seconds in *Aplysia* (Carew et al., 1981).

When conducting parametric studies of the CS–US interval across experimental groups, the effect of the interval can be made only on test trials in which the groups are exposed to the same relatively long CS (Bitterman, 1965). It is not appropriate to measure the effect of the CS–US interval on each trial, because the groups differ not only in terms of the CS–US interval but also in their opportunity to respond. It is incorrect to conclude that a group of crabs trained with a 1-second CS–US interval performs better than a group trained with a 5 second interval, unless both groups are tested with the same CS–US interval. In this case, we might test both groups with a 10-second interval.

Intertrial Interval

The intertrial interval refers to the period of time between CS presentations, or in the case of temporal conditioning, US presentations. In contrast to most invertebrate experiments (see Haralson et al., 1965 for an exception), vertebrate experiments typically vary the intertrial interval to guard against the possibility of temporal conditioning. In general, the shorter the intertrial interval is, the less the conditioning. There are few invertebrate studies that examine the influence of intertrial interval. Conditioning in intact earthworms, for example, is better with long intertrial intervals, whereas the reverse is true with decerebrate worms (Ratner, 1962; Ratner & Miller, 1959b). There seems to be little effect of long versus short intertrial intervals in proboscis conditioning of honeybees, although it does appear that short intervals (e.g., 1 min vs. 10 min) diminish performance somewhat. However, as training continues, performance is not as good as bees receiving long intertrial intervals (Bitterman et al., 1983).

Intertrial intervals vary widely in studies of invertebrate classical conditioning, and in general, they are usually much shorter than those used with vertebrates. For example, Abramson and Feinman (1988), in their studies of classical conditioning in the crab, used a variable intertrial interval averaging 1 minute. Smith and Baker (1960) also used a variable interval that ranged between 100 and 250 seconds in their studies of classical conditioning in horseshoe crabs. Lederhendler et al. (1986) used a variable intertrial interval in their studies of learning in *Hermissenda* that varied between 30 seconds and 4 minutes. Fixed intertrial intervals have ranged from 10 seconds in paramecia (Hennessey et al., 1979), 15 seconds with earthworms and planarians (Herz et al., 1964; Kimmel & Yaremko, 1966), 20 seconds in planarians (Thompson & McConnell, 1955), 30 seconds in the leech (Henderson & Strong, 1972), 50 seconds in earthworms (Ratner & Miller, 1959a), 1 minute in Nereids (Inozemtsev, 1990) 1 minute in houseflies, (Fukushi, 1979), 2.5 minutes in the leech (Sahley & Ready, 1988), 5 minutes in *Aplysia* (Carew et al., 1981), 10 minutes in houseflies (Fukushi, 1976), 10 minutes in bees (Bitterman et al., 1983), and 12–15 minutes in lobsters (Fine-Levy et al., 1988).

Some Phenomena of Classical Conditioning in Invertebrates

In this final section, we will take a brief look at some of the classical conditioning phenomena that have been examined in invertebrates.

Extinction and Spontaneous Recovery

Extinction refers to the gradual disappearance of the CR when the UR is removed. Following the passage of time, the CR often reappears. This is known as *spontaneous recovery*. The rate of extinction depends on many factors such as the number of training trials, intertrial interval, and the intensity of the CS and US. It is rather surprising that many demonstrations of classical conditioning of invertebrates do not incorporate an extinction session within the experimental design. According to research data, extinction is usually very rapid. For example, extinction of proboscis conditioning in honeybees and flies occurs within 6 presentations of the CS (Akahane & Amakawa, 1983; Bitterman et al., 1983); in the crab eye withdrawal reflex, 5 presentations (Abramson & Feinman, 1988); in the gill withdrawal reflex of *Aplysia*, 4 presentations (Lukowiak & Sahley, 1981); in the withdrawal responses of *Paramecium*, 8 presentations (Hennessey et al., 1979); in the withdrawal responses of earthworms, 20 presentations (Ratner & Miller, 1959a); in the withdrawal responses of leeches, 25 presentations (Henderson & Strong, 1972); and in the withdrawal responses of horseshoe crabs, 85 presentations (Smith & Baker, 1960). Spontaneous recovery curves are available for a number of invertebrates, including bees (Bitterman et al., 1983), leeches (Henderson & Strong, 1972), flies (Akahane & Amakawa, 1983), and *Paramecia* (Hennessey et al., 1979).

Generalization and Discrimination

Generalization refers to the tendency for the CR to be elicited by stimuli resembling the CS. The more similar the new stimulus is to the CS, the more vigorous is the CR. *Discrimination* is the opposite of generalization and refers to the ability to distinguish between stimuli. With the exception of the honeybee, both generalization and discrimination have received little attention in the invertebrate literature. Studies on generalization and discrimination in invertebrates are important not only for what they might reveal about the learning process but also what they show about sensory physiology (Getz & Smith, 1987; Smith, 1991).

The ability of invertebrates to discriminate between two stimuli—one of which is paired with a US—is often used as a pseudoconditioning control. This was done, for example, by Mpitsos and Cohan (1986) in their studies of classical conditioning in *Pleurobranchaea* by Fukushi (1979), Bitterman et al. (1983), and Fine-Levy et al. (1988) in their studies of olfactory conditioning in bees, flies, and lobsters, respectively, and by Block and McConnell (1967) in their studies of planarian conditioning.

Second-Order Conditioning

In the standard classical conditioning procedure, a CS is paired with a US. As the number of CS–US pairings increases, the CS begins to elicit a response that resembles the response elicited by the US (i.e., the CR). In *second-order conditioning*, one builds on this relation by using the original (*first-order*) CS as a US and pairs it with a new (second-order) CS. After a number of pairings between the new CS and the first-order CS (which now functions as a US), one begins to observe CRs to the new CS, even though this CS has never been paired with a biologically relevant US.

Second-order conditioning experiments with invertebrates are rare. Sahley, Rudy, and Gelperin (1981) were able to demonstrate that *Limax* can reduce its preference for a potato odor CS when this CS was paired with a first-order CS (carrot–quinine pairings). It should be noted, however, that the CSs used in this experiment were not neutral; taste aversion is not universally accepted as an example of classical conditioning; and, as Farley and Alkon point out (1985), poisoning may have produced an nonassociative neophobic reaction to any novel odor. Second-order conditioning was also reported in a proboscis conditioning experiment with honeybees (Bitterman et al., 1983).

Partial Reinforcement

Partial reinforcement refers to the periodic omission of the US. The general finding in such experiments, with the exception of fish and pigeons (Gonzalez, Longo, & Bitterman, 1961; Longo, Milstein, & Bitterman, 1962), is that partial reinforcement retards the acquisition of CRs. The effect of partial reinforcement has been investigated in the classical conditioning of the withdrawal response in earthworms. When the US was omitted on 50% of the trials, partial reinforcement was as effective as 100% reinforcement during acquisition. During extinction, however, partial reinforcement resulted in greater responsiveness than did consistent reinforcement (Wyers, Peeke, & Herz, 1964; Peeke, Herz, & Wyers, 1965). These results have been confirmed for the planarian (Kimmel & Yaremko, 1966).

Latent Inhibition

Latent inhibition refers to the retardation of excitatory conditioning by the unreinforced preexposure of a CS. Latent inhibition has been demonstrated in a variety of mammals (Lubow, 1973), but its status in non-mammalian vertebrates is unclear (Shishimi, 1985). In proboscis conditioning of honeybees, preexposure to the CS failed to retard acquisition (Bitterman et al., 1983). However, latent inhibition was found in a second honeybee experiment in which preexposure to vibration or air puff was found to retard the ability of bees to avoid shock in a free-flying situation (Abramson & Bitterman, 1986a).

A Checklist of the Essential Elements in a Study of Classical Conditioning

A list of the components that make up a classical conditioning experiment follows. The list will help you to organize and characterize invertebrate classical conditioning experiments.

1. Subject variables
   - Species comparison
   - Sex
   - Size
   - Age
   - Developmental stage
   - Number of animals, if any, discarded from the sample
   - Intact, free-behaving animal
   - Semi-intact or isolated preparation
   - Type of housing: isolated or group
   - Prior experience
   - Sensory capabilities
   - Neuroanatomical organization

2. Environmental variables
   - Characteristics of apparatus
   - Naturalistic versus laboratory environments
   - Temperature
   - Seasonal variability
   - Ecological manipulations

3. Type of conditioning
   - Excitatory
   - Inhibitory
   - Discrimination
   - Generalization
   - Compound

4. Response measures (intact, semi-intact, isolated preparations)
   - Frequency of response
   - Amplitude of response
   - Latency of response
   - Duration of response
   - Number of responses
   - Trials to reach criteria
   - Changes in topography

Individual differences
CR measured on each trial
CR measured on test trials
CRs measured during acquisition
CRs measured during extinction

5. Stimulus variables
CS intensity and duration
US intensity and duration
Partial or continuous reinforcement

6. Controls
Pseudoconditioning
Sensitization
Backward conditioning
Simultaneous conditioning
Differential conditioning
Base rate of responding
Temporal conditioning
Pheromones
CS only
US only
Central excitatory state (CES)
Systematic variation

Summary

In this chapter, I have discussed the similarities and differences between classical conditioning, habituation, and sensitization. In addition, we identified the necessary control procedures that must be used in a study on classical conditioning. The essential elements of a classical conditioning experiment were outlined.

Discussion Questions

• Compare and contrast classical conditioning with nonassociative learning.
• Describe the difference between alpha conditioning and classical

conditioning, and comment on whether you believe the distinction should be maintained.

- What are the similarities and differences between classical conditioning and instrumental and operant conditioning?
- What is the importance of classical conditioning in the life of an invertebrate? A vertebrate?
- What types of apparatus can be used in the study of classical conditioning?
- What are the essential elements in a study of classical conditioning?
- What experimental variables have been manipulated in studies of classical conditioning?
- Use the Guidelines for Planning or Reporting Experimentation to design an invertebrate study of classical conditioning.
- Use the Guidelines for Planning or Reporting Experimentation to find an experiment on classical conditioning in the library, and analyze it according to the guidelines.

6 Instrumental and Operant Conditioning

Preview Questions

- What are some general considerations in instrumental and operant conditioning?
- What are instrumental and operant conditioning?
- Why study instrumental and operant conditioning?
- What controls should be employed in a study of instrumental and operant conditioning?
- What are the essential elements in a study of instrumental and operant conditioning?
- What stimuli have been used in invertebrate experiments on instrumental and operant conditioning?
- What experimental variables have been manipulated in invertebrate experiments on instrumental and operant conditioning?

At the time that Ivan Pavlov was becoming interested in problems in behavior, Edward L. Thorndike (1911) published his doctoral thesis on animal learning describing what is now known as *instrumental* or *operant behavior* (see p. 39). In instrumental and operant conditioning, a contingency is arranged between a motivationally significant stimulus and a specific behavior. As a result of experience with the contingencies, the animal learns the consequences of its actions. For example, if an action is followed by a rewarding event such as food or the removal of pain, it tends to be repeated. On the other hand, if an action produces an aversive event, it tends not to be repeated.

There are four characteristics that differentiate instrumental and operant conditioning from classical conditioning. First, in instrumental and operant conditioning, an action produces reinforcement; in classical conditioning, an action should have no bearing on the receipt of reinforcement. Consider, for example, the difference between a lever-press situation and a Pavlovian dog. In the lever-press situation, an animal makes a response and moves to a location in the apparatus to collect its reward. The Pavlovian dog, however, is often given the reward by having it injected into its mouth. Second, the type of response that is learned in instrumental and operant conditioning is not restricted to the UR that is elicited by the reinforcement; in classical conditioning, it is. For example, in operant and instrumental conditioning, the type of response that you can train is quite broad and not directly tied to the reinforcer. Lever pressing, running to various locations, and making choices in mazes are just some of the behaviors that can be trained. In classical conditioning, however, the type of behaviors that you will train are often selected for you by the response that is elicited by the US. If the US elicits an eye blink, then the behavior you will train will involve an eye blink; if the US elicits proboscis extension in a fly, then the behavior that you will train will involve proboscis extension. Third, the experimenter generally has less control over the training variables in instrumental and operant conditioning than in a classical conditioning experiment. Recall from the previous discussion of nonassociative learning and of classical conditioning that the animal is essentially at the disposal of the experimenter. It is normally restrained and must remain in the experimental situation, where it is bombarded by a variety of stimuli.

In instrumental and operant conditioning, however, the animal can extract a little revenge. It can refuse to transverse a runway, decline to make a choice in a maze, or stop eating in a "Skinner box." If one is impatient, the animal can be hurried along by a swift "kick in the rear," or, as sometimes occurs, the animal can be unintentionally "guided" to the correct choice. To do so obviously invalidates the experiment. In classical conditioning, these problems are not normally encountered. Such precise control over experimental variables is a primary reason that those interested in the neuronal basis of learning have focused on classical conditioning, habituation, and sensitization rather than instrumental and operant conditioning. The fourth difference between classical conditioning and instrumental and operant conditioning is that there is no CS in operant conditioning to elicit a CR. Rather, there are stimuli known as *discriminative stimuli* that an animal can use as a cue in identifying a correlation between a response and reinforcement. The discriminative stimulus controls the instrumental or operant response because the response has been reinforced in its presence.

The presence of instrumental and operant behavior throughout the animal kingdom is less clear than that of nonassociative learning or classical conditioning. Basically, the problem is reduced to how one defines instrumental and operant behavior. If you use these terms interchangeably with "behavior controlled by its consequences," then such behavior is present in all animal groups. However, if operant behavior is defined in terms of its functional influence on the environment and the ability to use an arbitrary response, then the generality of operant behavior within the animal kingdom is limited to vertebrates and perhaps some species of molluscs, crustaceans, and insects. A rule of thumb that I have found useful to distinguish between instrumental and operant behavior is that in operant behavior, an animal must demonstrate the ability to *operate* some device—and *know* how to use it, that is, make an *arbitrary* response to obtain reinforcement. The term *instrumental behavior* would therefore be reserved for the response contingent modification of responses that are not arbitrary.

What Are Instrumental and Operant Conditioning?

As mentioned earlier, instrumental and operant conditioning are examples of associative learning in which the behavior of the animal is controlled by the consequences of its actions. As we saw in chapter 3, instrumental and operant conditioning can be studied using a variety of apparatus such as running wheels, runways, mazes, shuttle boxes, leg-lift learning, and lever-press boxes. Instrumental and operant conditioning are generally thought to be more complex than classical conditioning. One might roughly characterize the difference by saying that classical conditioning describes how animals make associations between stimuli, and instrumental and operant conditioning describe how animals associate stimuli with their own motor actions. Classical conditioning emphasizes sensory integration, and instrumental and operant conditioning, motivation: Animals learn new behaviors in order to obtain or avoid some stimulus (reinforcement). In addition, instrumental and operant conditioning are thought to be more complex than classical conditioning because learning depends on the animal's own behavior and usually requires a more obviously new behavior. Despite these differences, instrumental and operant conditioning share many properties with classical conditioning. These include extinction, spontaneous recovery, generalization, and discrimination.

Perhaps the most familiar images of instrumental and operant conditioning are the maze and the "Skinner box." Everyday examples of such conditioning are so numerous that it is easy to forget that much of human behavior is controlled by its consequences. Most people, for instance, can readily identify acts performed to maximize pleasure and minimize pain. This is true also at the invertebrate level. Abby-Kalio (1989) has shown that the shell-cracking behavior of the green crab improves as a result of instrumental training, as does the learning of predatory skills (Cunningham & Hughes, 1984).

In contrast to classical conditioning, the definition of instrumental and operant behavior is generally consistent from experimenter to experimenter. If we were to once again reconvene our group of psychologists, zoologists, and physiologists and ask them to define instrumental and operant conditioning, most of them would say that it is the control of behavior by its consequences and that you can use either term interchangeably. There will be a small percentage of psychologists, however, that would maintain that there is a difference between instrumental and operant behavior.

In this book I will make a distinction between operant and instrumental conditioning. The reason for distinguishing instrumental and operant conditioning is not necessarily because different principles are involved. In each case, behavior is controlled by its consequences. You will often find, in the contemporary literature on learning, that the terms *instrumental conditioning* and *operant conditioning* are currently used interchangeably to identify procedures in which reinforcement is contingent upon an act. There are some differences, however, that you should consider. Operant behavior is studied in situations such as the lever-press box described in chapter 3, in which an animal can respond to the contingencies of the experiment without interruption. In addition, an operant response leaves the animal in basically the same position in space; you might say it leaves the animal "buried in space." Instrumental behavior, on the other hand, is studied in runways and mazes in which the animal moves a relatively long distance. In other words, it is "lost in space." There are also abrupt breaks in the behavior stream. These abrupt breaks are known as *discrete trials*.

A second difference between operant and instrumental techniques is that operant procedures minimize the animal's "species-typical" behavior in order to facilitate the functional analysis of behavior. This is the reason that manipuladums such as levers and treadles are used. In addition, such responses have the property of being performed effortlessly for long periods of time. For instance, a crab can effortlessly press a lever, and the data can be compared to the lever press of a bee, snail, and *Aplysia* in terms of the parameters known to influence operant behavior (e.g., schedule of reinforcement, behavioral contrast). Such a comparison is appropriate despite the fact that there are behavioral and physiological differences between crustaceans, insects, snails, and molluscs. Of course, if you are interested in species differences in response topography, then the lever-press technique should not be used.

Another important difference between instrumental and operant procedures is the response measure used as the index of learning. Typically, operant experiments emphasize the rate at which the response is made, and instrumental experiments emphasize the latency or amplitude of a response. Although response rate is a stable and sensitive measure of behavior, in some cases it is not as useful as the latency or amplitude measure. This is particularly true if you are interested in the acquisition of a response.

In operant experiments, the study of acquisition is problematic because the experimenter cannot control, at least in the case of positive reinforcers, when the animal actually consumes the reward. Moreover, the URs associated with eating or drinking get in the way of the analysis of acquisition. In instrumental experiments, this is controlled by confining consummatory activity to a separate portion of the apparatus, such as the goal box of a runway or maze. In this way you can, for example, separate the speed of an ant as it runs down an alley (speed is the acquisition measure) from the time it takes to drink from a moist sponge (water is the reinforcer and drinking is the consummatory response).

An additional complication in operant experiments is that the animal determines the intertrial interval: It is free to respond at any moment. As we mentioned in the section on classical conditioning, the level and ease of acquiring a learned response is partly determined by this interval. In instrumental conditioning, the intertrial interval is easily manipulated by restricting the animal to the start box and releasing it at the appropriate time.

Two other differences between instrumental and operant conditioning deserve mention. First, operant experiments typically employ few subjects and study them over a period of time that is usually longer than that used in instrumental situations. It is not uncommon to conduct an operant experiment in which only four animals are studied over several months. Second, it is often difficult in operant situations to study learning with aversive reinforcers such as shock because such stimuli elicit behavior that interferes with the operant response. For example, if you shock a crab every time it presses a lever, you soon have an "armless" crab. Such a condition is not conducive to lever pressing. In the instrumental situation, this is normally not a problem. If your interest lies in aversive conditioning of invertebrates, it would be wise to consider using instrumental or classical conditioning techniques.

It should be emphasized that free-operant and instrumental discrete trial techniques are equally effective in revealing lawful properties in behavior. Moreover, each technique shares the trait of reciprocity. In other words, discrete trial techniques can be woven into free-operant designs, for instance, by having a lever inserted into an operant chamber to signal the start of a trial. Alternatively, free-operant techniques can be incorporated into instrumental designs, as when the runway is adapted for the study of reinforcement schedules (Logan, 1960). Instead of, for example, having a crab press four times to obtain a piece of squid, simply have it run down an alley four times. In both cases, the animal must make four responses to obtain the reward. The response measures can also be woven together. The latency of response can be adapted to the operant chamber by signaling the start of a trial and recording the time taken to lever press. If discriminative stimuli are not available, latency can be recorded by using a retractable lever that enters the chamber at the start of a trial. Rate of response can be recorded in a runway by measuring the speed of the animal as it passes successive points.

Why Study Instrumental and Operant Conditioning in Invertebrates?

The reasons for studying instrumental and operant conditioning in invertebrates are the same as those discussed for the study of classical conditioning. First, the study of instrumental and operant conditioning has value as a comparative tool in testing the generality of psychological and comparative theories of behavior and physiology. You must be aware, however, that for some classes of invertebrates, instrumental conditioning may be only a laboratory phenomenon. Rosenkoetter and Boice (1975) suggest, for instance, that earthworms have little use for associative learning in nature, and Dethier (1966) suggests that some insects have lost the ability to learn. Pheromones and nonassociative learning mechanisms may be all that is necessary to produce adaptive changes in behavior.

A second reason for studying instrumental and operant conditioning is that these powerful techniques can be used in sensory physiology in determining sensory thresholds and uncovering the underlying neuronal mechanisms of olfaction, taste, vision, and hearing (Kirchner, Dreller, & Towne, 1991). These techniques can also be applied in the field of neuronal learning models. Compared to nonassociative and classical conditioning, there are few neuronal models of instrumental and operant conditioning. The notable exceptions are the head-waving response in *Aplysia* (Cook & Carew, 1986) and leg position in the locust (Forman & Zill, 1984; Hoyle, 1980) and crab (Dunn & Barnes, 1981a, 1981b; Punzo, 1983).

The study of instrumental and operant conditioning is also important because it offers an opportunity to understand at the neuronal level the similarities and differences between various types of behavior. As was discussed in the last chapter, there is no accepted classification of classical conditioning; the same is true for instrumental and operant conditioning. Perhaps the problem of classifying behavior can be made at the neuronal level.

The study of instrumental and operant conditioning at the invertebrate level is also an interesting design problem. In comparison to vertebrate studies, most of the available instrumental and operant techniques are rather crude. This is especially obvious when compared to vertebrate techniques and even to invertebrate studies of classical conditioning and nonassociative learning.

What Controls Should Be Used?

Before it can be concluded that a change in behavior is the result of the formation of an association between a response and a reinforcer, several alternative explanations must be eliminated. The problem of control is not as great as it is in classical conditioning because in principle, the instrumental and operant response should be a behavior that is not in the repertoire of control animals. For example, if several crabs are trained to press a lever for food and then placed in a general population of untrained crabs, it should be an easy matter to determine the trained crabs from the untrained by placing them in the operant chamber. In addition, many maze experiments have a built-in control by requiring the animal to make a discrimination between two stimuli.

Pseudoconditioning

As we saw in the previous chapter on classical conditioning, one of the factors that must be guarded against is that US presentation can affect the animal so that a subsequently presented CS produces a "CR," even though the CS and US have never been paired. Pseudoconditioning is another pitfall in instrumental and operant experiments (see p. 129). In this case, it must be ensured that the reinforcement does not produce a change in behavior resembling the index of instrumental or operant conditioning. As an example, let us consider a study of the acquisition of runway performance in an ant with access to water as the reward. Over a series of 20 trials, the speed of the ant is observed to increase with training. Can we conclude that the increased speed represents that the thirsty ant has associated the goal box with water? The answer is no. It is just as likely that the increased speed would have occurred without any reward in the goal box. Increased speed might also represent the energizing effect of exposure to water that is independent of the contingency between the response and the reward. To assess the amount of pseudoconditioning, a control group is necessary that occasionally receives water in a location other than the goal box. It is also possible to assess pseudoconditioning by training the ant to discriminate between two stimuli, one of which leads to water in the goal box.

Pseudoconditioning must also be assessed in operant lever-press situations. Let us consider an example in which a crab is trained to press a lever for food. The animal is placed in a small operant chamber, and each bar press is rewarded with a bit of squid. Following the first few bar presses, one can soon observe that the rate of pressing increases dramatically. Can we conclude that the increase in lever pressing is the result of the contingency between lever press and food? The answer is once again no. It is entirely possible that the increase is due to the energizing influence of food. The more active the animal, the greater the probability that it will contact the bar. What is needed in this situation is for a second animal to receive the same amount of food as the experimental animal, but the food must be independent of the lever pressing. This type of control is called a *yoked control*.

A second way to assess pseudoconditioning is to train the crab to discriminate between two levers, one of which is reinforced by food. Another is by using one lever, which the animal is trained to press only in the presence of a discriminative stimulus such as light. It is also possible to use a third method in which some property of the response, such as its speed, latency, or rate of responding is differentially rewarded. One might train the ant to run down the alley within a certain time period or train the crab to press a lever with a certain degree of force. Although the latter methods are standard procedures with vertebrates, one can imagine the difficulty of applying them to invertebrates. In the case of lever pressing, one can get around this problem by creating a behavior that is not in the "vocabulary" of control animals. This is known as *shaping*. With the proper use of reward and extinction, one should be able to train a crab, for example, to press a bar that cannot be reached unless postural adjustments are made.

It should be mentioned that the yoked control has been criticized in both the vertebrate and invertebrate literature and should be used with caution. Church (1964) has pointed out that because of the nature of the yoked design, animals with low reactivity to the stimulus will perform only at a baseline level (they cannot do any worse), and therefore, because animals with a high reactivity will appear to improve, a random difference in inherent responsiveness will lead to artifactual learning in the population. This artifact cannot be eliminated by increasing the sample size. Church showed that if there is an inherent difference in reactivity to the reinforcement, 25% of the animals will have appeared to "learn." Church and Lerner (1976) have applied this analysis directly to the case of leg lift learning in the roach (the Horridge paradigm), and by making simple assumptions for inherent reaction to aversive stimuli, a random model predicted that controls would maintain the punished position up to 50% of the time. This analysis has been confirmed by Buerger, Eisenstein, and Reep (1981). These authors maintain, however, that temporal regularities in the performance of Horridge-trained roaches are consistent with a learning interpretation. Church has suggested that the yoked control be used in conjunction with other controls, such as introducing a delay between the response and reinforcement. An example of this strategy can be found in an application of the Horridge paradigm to the eye withdrawal reflex in the crab (Abramson & Feinman, 1987).

Base Rate of Responding

As in nonassociative learning and classical conditioning, it is important to know the frequency of the behavior that is to serve as the index of learning. In operant experiments, such a record is known as the *operant* or *base rate of responding*. As mentioned earlier, such a control group is especially important in runway studies because it is likely that an increase in running speed of an invertebrate will occur independently of the re-ponse–reinforcer relationship. This sort of control group is also useful as an indicator of the influence of calendar variables.

Position Preferences

When conducting experiments in which the invertebrate is confronted with an alternative between two levers or two or more choice points, it is important to determine whether one side or the other is favored by the animal before testing. Such bias is well known in invertebrate research and can be caused by a number of factors, such as imperfections in the apparatus, the route taken by animals as they approach choice points, differences in the brightness between discriminative stimuli, and color and odor preferences. Controlling for such preferences is particularly important in spatial discrimination tasks in which the invertebrate is trained to go always in one direction. Position preferences can be assessed by running a reference group to determine if such biases exist. If they do, the experimental animals are run against this preference. In some cases, it is necessary that the experimental group be given a preference test prior to training, and based on this test, that animals be trained to go against the preference.

In visual discrimination tasks, it is also important to determine whether the stimuli you select introduce biases. It is common practice in visual discrimination experiments to counterbalance the position of the stimuli. For example, when free-flying honeybees are trained to discriminate be-tween two targets differing in color, the position of the two stimuli are varied according to some pseudo-random order.

Calendar Variables

The influences of seasonal variables and changes in local weather con-ditions have a strong influence on invertebrate behavior. Although these variables are difficult to manipulate, they can be controlled for by inter-mingling the running of the experimental and control groups. Thus, if you have 20 experimental roaches and 20 control roaches, do not run all 20 experimental roaches before running the control animals. Rather, for every experimental animal, run a control animal.

Pheromones

The influence of pheromones in the maze and runway performance of invertebrates has been extensively investigated. Their presence will influence performance. This has been shown, for instance, in maze studies of the earthworm (Rosenkoetter & Boice, 1975). It is common practice to control for such odors by cleaning the apparatus before each trial and for each new subject. Unfortunately, in some cases, cleaning the apparatus may not be enough. Ressler, Cialdini, Ghoca, and Kleist (1968) have reported that earthworm alarm pheromone is extremely persistent and not easily removed.

Systematic Variation

When comparing the results of instrumental and operant experiments across species, it is important to keep in mind that any differences you encounter may be the result of differences in sensory abilities, motor demands, motivational levels, or reward values. (The same is true in studies of classical conditioning and nonassociative learning.) Let us consider a situation in which each lever press of a crab and of *Aplysia* is followed by food. When reward is discontinued after 50 presses and we find that the crab emits more responses than the mollusc, can we say that the crab is more resistant to extinction? The answer is no. It is possible that differences can be accounted for in terms of deprivation level. The crab may have been initially more hungry than the *Aplysia*. This question can be answered by what Bitterman (1965) has called control by systematic variation. If the greater persistence of lever-press behavior in the crab is due to differences in motivational level, there should be a level of deprivation at which the *Aplysia* would be more persistent than the crab. It becomes a relatively simple matter to test this idea by conducting a second experiment in which deprivation level is manipulated in *Aplysia*.

Experimenter Bias

It is just as important to control for experimenter bias in instrumental and operant conditioning as it is in the study of classical and nonassociative learning. If it is possible to automate your experiment, attempt to do so. If it is not possible, videotaping your experiment will ensure that you are not measuring your skill in pushing an animal down a runway or maze.

What Are the Essential Elements in a Study of Instrumental and Operant Conditioning?

In this section we will take a look at some of the basic conditioning paradigms used in the study of instrumental and operant conditioning.

The basic instrumental and operant conditioning paradigms are classified according to the type of reinforcement used (desirable or undesirable) and whether the response produces a reinforcement that is desirable or undesirable. This classification produces four basic conditioning paradigms that you have at your disposal to manipulate the consequences of an invertebrate's behavior. First, if the instrumental or operant response produces a desirable reinforcer—such as a crab pressing a lever for squid or for the termination of shock—the paradigm is reward and escape training, respectively. As Bitterman (1979) notes, escape training is a special case of reward training in which the desirable event is the termination of an aversive stimulus. Second, if the response produces an undesirable event such as shock, the paradigm is punishment training. Third, if the response prevents the delivery of a desirable event, the paradigm is omission training. The fourth category is a special case of omission training in which the response prevents the delivery of an aversive event. This is known as *signaled avoidance*.

In operant conditioning, there is a second type of avoidance situation known as *unsignaled* or *Sidman avoidance*, named after the originator of the procedure, Murray Sidman. The difference between them is that in signaled avoidance, a warning stimulus signals the presentation of the aversive stimulus; in unsignaled avoidance, there is no such signal. Rather, the passage of time serves as a signal and a response postpones the scheduled aversive event.

Reward Training

Reward training is one of the most common instrumental and operant procedures for the study of invertebrate learning. Such training with invertebrates is usually carried out in maze and runway situations. With the ascendancy of the honeybee as the favorite experimental animal in reward experiments, most reward experiments are carried out with the free-flying honeybee procedure. Several different types of discriminative stimuli have been used, including odor (Couvillon & Bitterman, 1980; Vowles, 1964), color (Couvillon & Bitterman, 1980; Sigurdson, 1981a), and shapes (Vowles, 1965).

Various types of rewards have been found to be effective with invertebrates. The most common are food, water, escape from aversive stimulation, and return to a home container. The retrieval of pupae has been found to be a very effective reinforcer for ants (Dashevskii, Karas,

& Udalova, 1990). The use of pupae has the added advantage that motivation for this reward is more consistent than that for food or water.

A reward that has not received much attention in the invertebrate literature but that should be highly effective is the removal of aversive stimulation in escape training. Escape training has seldom been investigated with invertebrates, but there is much to recommend it. Motivation and deprivation level can be controlled to a greater degree than is generally possible in food- or water-deprived invertebrates, and both the amount of reduction in some aversive event and the length of the "aversive free time" produced by the escape response can easily be manipulated. Moreover, as Mackintosh (1983) argued on the basis of the vertebrate literature, there is an essential similarity between appetitive and aversive conditioning. Escape training may be an ideal technique to investigate learning in a wide variety of invertebrates that would otherwise be ignored because of the difficulty in finding suitable food or water rewards. Aversive stimuli that have been shown to be effective for invertebrates include shock, heat, odor, and alarm pheromones.

Punishment Training

A second procedure that has found widespread use with invertebrates is the punishment paradigm. At one time, invertebrate punishment experiments were common; now they are relatively rare. Punishment was used often in invertebrate maze experiments to facilitate the selection of a correct choice, as, for instance, when an earthworm was shocked for an incorrect choice (Datta et al., 1960). More often, however, punishment was investigated in the Horridge paradigm in which precise movements of an appendage were punished (see chapter 3 for several examples). More recently, punishment has been adapted to the free-flying bee situation as well as to the proboscis conditioning situation (Abramson, 1986; Smith, Abramson, & Tobin, 1991). The most common aversive stimulus used in punishment experiments is electric shock.

There have been only a few examples in which discriminative stimuli are employed in punishment experiments with invertebrates. Bees have been trained to discriminate between two odors or differently colored targets on the basis of punishment, even though both stimuli are associated with reward (Abramson, 1986; Smith, Abramson, & Tobin, 1991).

Omission Training

Studies on omission training are rarely performed with invertebrates. Bitterman et al. (1983) used the procedure to examine the role of ad-

Table 6-1

Samples of Types of Tasks and Rewards Used With Invertebrates

Animal	Task	Reinforcer	Reference
		Reward training	
Ant	Maze	Food	Schneirla (1933)
	Maze	Home container	Vowles (1965)
	Maze	Pupae	Dashevskii et al. (1990)
	Maze	Water	Simmel & Ramos (1965)
	Escape	Peppermint odor	Hoagland (1931)
	Escape	X ray	Martinsen & Kimeldorf (1972)
	Escape	Water	Morgan (1981)
Aplysia	Operant	Cooling	Downey & Jahan-Parwar (1972)
Bee	Maze	Food	Kirchner et al. (1991)
	Free-flying	Food	Couvillon & Bitterman (1980)
	Operant	Food	Sigurdson (1981a, 1981b)
	Operant	Food	Grossmann (1973)
	Operant	Food	Pessotti (1972)
	Escape	Formic acid odor	Abramson (1986)
	Escape	Shock	Abramson (1986)
Beetle	Maze	Home container	Alloway (1969)
Crab	Maze	Food	Yerkes (1902)
	Shuttle box	Food	Karas (1962)
	Operant	Food	Abramson & Feinman (1990a, 1990b)
Crayfish	Maze	Low pH	France (1985)
	Operant	Food	Olson & Strandberg (1979)
Fly	Maze	Ascent	Platt, Holliday, & Drudge (1980)
Fly	Escape	Light/heat	Leeming & Little (1977)
Fly	Foraging	Food	Fukushi (1989)
Planarian	Maze	Home container	Corning (1964)
Roach	Runway	Food	Longo (1970)
	Operant	Food	Rubadeau & Conrad (1963)
Snail	Operant	Brain stimulation	Balaban & Chase (1989)
		Punishment training	
Ant	Shuttle box	Vibration	Abramson (1981)
Aplysia	Head waving	Light	Cook & Carew (1986)
Bee	Shuttle box	Formic acid odor	Abramson (1986)
	Proboscis extension	Shock	Smith et al. (1991)
	Free-flying	Shock	Abramson (1986)
	Lever-press	Shock	Pessotti & Lignelli-Otero (1981)
Crab	Maze	Shock	Cuadras, Vila, & Balasch (1978)
	Respirometer	Hypoxia	Becker & Valinski (1981)
	Horridge	Air puff	Abramson & Feinman (1987)
	Horridge	Shock	Punzo (1983)
	Shuttle box	Shock	Denti, Dimant, & Maldonado (1988)

continues

Table 6-1, Continued

Animal	Task	Reinforcer	Reference
		Punishment training	
Crayfish	Horridge	Shock	Stafstrom & Gerstein (1977)
Earthworm	Runway	Sodium chloride	Wyers, Smith, & Dinkes (1974)
Fly	Horridge	Shock	Booker & Quinn (1981)
Locust	Horridge	Heat	Forman (1984)
Snail	Horridge	Shock	Emson, Walker, & Kerkut (1971)
	Horridge	Shock	Christoffersen et al. (1981)
Roach	Horridge	Shock	Harris (1991)
	Learned helplessness	Shock	Brown, Hughes, & Jones (1988); Brown & Stroup (1988)
		Omission training	
Bee	Proboscis extension	Food	Bitterman et al. (1983)
	Operant	Food	Sigurdson (1981a)
		Reversal/probability learning	
Bee	Choice	Food	Sigurdson (1981a)
Crab	Maze	Preferred environment	Datta, Milstein, & Bitterman (1960)
Crayfish	Maze	Preferred environment	Costanzo & Cox (1971); Costanzo, Rudolph, & Cox (1972)
Earthworm	Maze	Preferred environment	Datta (1962)
Isopod	Maze	Preferred environment	Harless (1967)
Roach	Maze	Preferred environment	Longo (1964)
Sow bug	Maze	Preferred environment	McDaniel (1969)
		Avoidance training: Signaled	
Bee	Free-flying	Shock	Abramson (1986)
Crab	Withdrawal	Vibration	Abramson et al. (1988)
Crayfish	Shuttle box	Shock	Taylor (1971)
Earthworm	Runway	Shock	Kirk & Thompson (1967)[a]
	Movement	Light	Ray (1968)
Fly	Shuttle box	Heat	Leeming (1977)[a]
Horseshoe crab	Withdrawal	Shock	Makous (1969)[a]
	Withdrawal	Shock	Wasserman & Patton (1969)
Locust	Foraging	Food	Bernays & Wrubel (1985)
Planarian	Withdrawal	Shock	Ragland & Ragland (1965)
Roach	Leg lift	Shock	Chen et al. (1970)
	Leg lift	Shock	Pritchatt (1970)[a]
		Avoidance training: Unsignaled	
Bee	Shuttle box	Formic acid odor	Abramson (1986)[b]
Roach	Maze	Shock	Longo (1964)[b]

Note. [a]Negative results. [b]No evidence of temporal regularity.

ventitious response–reinforcer contiguity in proboscis conditioning in honeybees, and Sigurdson (1981a) used the procedure in his studies of instrumental conditioning in honeybees.

Avoidance Training

Although not as rare as studies of omission training, there are few studies of avoidance in invertebrates. Signaled avoidance is unique because it is composed of elements from both classical and instrumental conditioning. A warning stimulus is paired with an aversive event in much the same way that a CS is paired with a US. The interesting twist is that the pairing is not entirely independent of the animal. When the animal makes a response to the warning stimulus, the aversive event is no longer scheduled to occur. Signaled avoidance has been demonstrated in honeybees (Abramson, 1986) and green crabs (Abramson, Armstrong, Feinman, & Feinman, 1988). It has been less convincingly demonstrated—because of lack of control procedures—in roaches (Chen, Aranda, & Luco, 1970), earthworms (Ray, 1968), planarians (Ragland & Ragland, 1965), and crayfish (Taylor, 1971). Negative results have been obtained with houseflies (Leeming, 1985), horseshoe crabs (Makous, 1969), roaches (Pritchatt, 1970), and earthworms (Kirk & Thompson, 1967). Unsignaled avoidance has been demonstrated in honeybees (Abramson, 1986) and roaches (Longo, 1964). There was no evidence, however, of the existence of temporal discrimination that is so common in vertebrate studies of unsignaled avoidance. The type of warning stimuli used in signaled avoidance studies range from vibration (Abramson, Armstrong, Feinman, & Feinman, 1988; Ray, 1968), to light (Makous, 1969; Pritchatt, 1970; Taylor, 1971), to shock (Chen, Aranda, & Luco, 1970). Table 6-1 presents a representative sample of the type of tasks and rewards that have been used with invertebrates.

Some Phenomena of Instrumental and Operant Conditioning in Invertebrates

In this section we will take a brief look at some of the phenomena of instrumental and operant conditioning that have been examined in invertebrates.

Reward Conditioning

Amount of Reinforcement

The amount or magnitude of reinforcement is one of the key variables influencing instrumental and operant conditioning. The vertebrate data generally indicate that both the rate of acquisition and the terminal level of performance are positively related to reward magnitude. Moreover, there is some suggestion in the literature that vertebrates learn "about" the reward. In other words, they form a representation of the reward. There are only a few invertebrate investigations on the influence of the amount or magnitude of reinforcement. The notable exception is work with the honeybee that suggests learning in the free-flying situation is a function of sucrose concentration (see Buchanan & Bitterman, 1988, 1989; Couvillon, Lee, & Bitterman, 1991; Lee & Bitterman, 1990a; Lee & Bitterman, 1992). There are no invertebrate studies that manipulate the amount of reinforcement in escape paradigms.

Delay of Reinforcement

The degree to which reinforcement is delayed is also a critical variable in reward conditioning, analogous to the CS–US interval in classical conditioning. The vertebrate data suggest that in both reward and punishment conditioning, performance is inversely related to reinforcement delay. Although there are few studies, the invertebrate data are in agreement. The majority of these studies can be found in the punishment literature. Using modified versions of the Horridge paradigm, both Mariath (1985) using *Drosophila* and Abramson and Feinman (1987) using crabs found poorer performance with longer delay between the response and onset of aversive stimulation. In addition, a 5-second delay between feeding and shock produced less suppression of feeding than when there was no delay in a free-flying honeybee situation (Abramson, 1986). Delay of reinforcement also influences performance in appetitive situations (Couvillon & Bitterman, 1980; Lee & Bitterman, 1990b).

Contrast Effects

Contrast refers to how shifts in the amount or quality of reward influence performance. The study of contrast holds an important place in the history of learning because it was research using contrast experiments that first suggested a distinction between learning and performance. With the exception of honeybees, there has been no attempt to study contrast in invertebrates. Successive negative contrast, incentive contrast, and behavioral contrast have been discovered using the free-flying procedure (Bitterman, 1976; Couvillon & Bitterman, 1984; Sigurdson, 1981a).

Partial Reinforcement

Another variable that is known to influence performance in instrumental conditioning is the percentage of the total responses that are rewarded or punished. You may recall that this variable is also important in classical conditioning. In contrast to classical conditioning, the vertebrate data suggest that there is usually no impairment in the acquisition of an instrumental response. When reinforcement is withheld during extinction, however, both classically conditioned and instrumentally conditioned responses show greater persistence.

The study of partial reinforcement is important for what it reveals about the influence of reward and nonreward on performance. Moreover, a case has been made by those who study operant conditioning that partial reinforcement approximates more closely events in the environment. Few of us outside the laboratory, for example, receive a reward for each occurrence of an appropriate behavior, yet such behavior does not decline.

The effect of partial reinforcement has been investigated in both instrumental runway and maze situations as well as operant situations. A principal difference between the procedures is that operant studies typically examine the effect of partial reinforcement on behavior that has been well practiced, whereas instrumental situations are concerned more with the effect on acquisition.

There are few studies on partial reinforcement in invertebrates. Recently, Ishida, Couvillon, and Bitterman (1992), using the free-flying honeybee procedure, have shown that resistance to extinction is increased, based on the pattern of reinforcement given during acquisition. Most studies of partial reinforcement in invertebrates have been performed using operant lever-press or hole-dipping procedures.

There are several ways to program the pattern and frequency of reward in operant experiments. The most basic are known collectively as *simple schedules* of reinforcement. They are *fixed ratio* (FR), *fixed interval* (FI), *variable ratio* (VR), and *variable interval* (VI). These basic schedules can be combined to form what is known as *complex schedules*. For example, a *multiple schedule* is composed of two or more independent schedules, each with its own discriminative stimulus. A *mixed schedule* is identical to a multiple schedule, with the exception that there are no discriminative stimuli. The animal may also be asked to satisfy the requirements of two or more schedules, as when, for example, it must emit 30 responses within

1 minute. This is known as a *compound schedule*. Each type of schedule produces a characteristic pattern of responding. There are several invertebrate studies that investigated responding on simple schedules (i.e., FI, VI, FR, VR), but no studies employing multiple or compound schedules.

Despite the number of operant procedures available with invertebrates (see Table 6-1), there are few investigations of partial reinforcement or of the many variables that have been found to influence performance, such as punishment and amount of reinforcement. Following are the available operant studies pertaining to the effect of partial reinforcement:

Fixed Ratio. In a FR schedule, reinforcement is available following a fixed number of responses. Using dipping into a tube as the operant response, both Grossmann (1973) and Sigurdson (1981a) demonstrated that bees that are trained on FR are more persistent during extinction than animals that receive reinforcement for each response. The number of dipping responses that are required to obtain reinforcement has been as high as 30. Crabs can be trained to press a lever when the response requirement is 9 presses to obtain reinforcement. The effect is not very strong, however, and there are data on resistance to extinction (Feinman, Korthals-Altes, Kingston, Abramson, & Forman, 1990).

Fixed Interval. In an FI schedule, reinforcement is available after a fixed period of time has elapsed. With extended training, animals begin to discriminate time. They respond less after reinforcement and more as the time for reinforcement approaches. Fixed interval schedules have been investigated by Grossmann (1973) and Sigurdson (1981a) for bees and Abramson (unpublished observations) for crabs. In each of these experiments, the invertebrate persisted in responding, but there was no evidence of the temporal regularity in performance that so characterizes higher vertebrates. Grossmann (1973), for example, required the bee to wait 90 seconds for a sucrose reward. Although the bee could do this, Grossmann remarked that no temporal regularity developed. A lack of temporal regularity appeared also in a study of the lever-press behavior in the crabs that were given extensive training on FI at 5-, 10-, and 15-second schedules.

Variable Ratio. In a VR schedule, the number of responses required for reward vary from one reinforcer to the next in an irregular manner. There are no invertebrate studies of variable ratio schedules.

Variable Interval. In a VI schedule, the time required for reward varies from one interval to the next in an irregular manner. There are no invertebrate studies of variable interval schedules.

A Checklist of the Essential Elements in a Study of Instrumental and Operant Conditioning

In this section the components that make up an instrumental and operant experiment are listed. The checklist will help you to organize and characterize instrumental and operant conditioning experiments.

1. Subject variables
    Species comparison
    Sex
    Size
    Age
    Developmental stage
    Number of animals, if any, discarded from the sample
    Intact, freely behaving animal
    Semi-intact or isolated preparation
    Type of housing: isolated or group
    Prior experience
    Sensory capabilities
    Neuroanatomical organization
2. Environmental variables
    Characteristics of apparatus
    Naturalistic vs. laboratory environments
    Temperature
    Seasonal variability
    Ecological manipulations
3. Type of conditioning
    Instrumental
    Operant (Is the response arbitrary?)
    Escape
    Reward
    Omission
    Avoidance
4. Schedule of reward
    Continuous
    Intermittent
        Fixed ratio
        Fixed interval
        Variable interval
        Variable ratio

        Mixed
        Multiple
        Concurrent
        Pattern
     Number of training trials
     Intertrial interval

5. Response measures (intact, semi-intact, isolated preparations)
     Frequency of response
     Amplitude of response
     Latency of response
     Duration of response
     Number of responses
     Trials to reach criteria
     Changes in topography
     Individual differences
     Rate of response
     Instrumental or operant response measured during acquisition
     Instrumental or operant response measured during extinction
       acquisition

6. Stimulus variables
     Intensity, duration, and magnitude of reward
     Delay of reward
     Type of reward

7. Controls
     Pseudoconditioning
     Base rate of responding
     Position preferences
     Calendar variables
     Pheromones
     Systematic variation
     Experimenter bias

Summary

In this chapter we have looked at the similarities and differences between the three major types of associative learning: Classical conditioning, instrumental conditioning, and operant conditioning. In addition, it was suggested that a distinction be made between instrumental conditioning and operant conditioning. The control procedures and essential elements necessary to conduct an experiment on instrumental and operant conditioning were identified and discussed.

Discussion Questions

- Compare and contrast instrumental and operant conditioning.
- What are the similarities and differences between classical conditioning, instrumental conditioning, and operant conditioning?
- Should the distinction between instrumental and operant conditioning be maintained?
- What is the importance of instrumental and operant conditioning in the life of an invertebrate? Of a vertebrate?
- What type of apparatus can be used in the study of instrumental and operant conditioning?
- What are the essential elements in a study of instrumental and operant conditioning?
- What experimental variables have been manipulated in studies of instrumental and operant conditioning?
- Use the Guidelines for Planning or Reporting Experimentation to design an invertebrate study of instrumental and operant conditioning.
- Use the Guidelines for Planning or Reporting Experimentation to find an experiment on both instrumental and operant conditioning in the library, and analyze it according to the guidelines.

The Cellular Analysis of Invertebrate Learning

Preview Questions

- Why are invertebrates useful for the cellular analysis of learning?
- What are some trends in the evolution of nervous systems?
- What are the major strategies in a cellular analysis?
- What are the major cellular models of habituation and sensitization?
- What are the major cellular models of classical and operant conditioning?

A major use of invertebrate animals in psychology and related disciplines in the neurosciences is to explain the biophysical and biochemical changes that occur in the nervous system during learning. As discussed in the first two chapters, invertebrates are particularly well suited to use in research on the neural mechanisms of learning. In this chapter, I will focus on the strategies that you will need to consider in conducting a cellular analysis. The data obtained with this analysis have been amply reviewed, and I will therefore touch the data for illustrative purposes only. If you are interested in a detailed analysis, I urge you to consult the reference material in Exhibit 7-1.

An implicit rationale for believing that the study of invertebrate nervous systems can reveal something about the way you and I function is the view that human behavior and that of an invertebrate can ultimately be explained by an understanding of molecular events. The function of the human senses—and of those of an invertebrate—is after all to translate the environment into a molecular "language" that the respective

Exhibit 7-1

General Issue-Oriented and Review Articles

1. Important issue-oriented articles
   The articles that follow are a must-read.
   > Corning & Lahue (1972)
   > Abraham, Palka, Peeke, & Willows (1972)
   > Corning, Dyal & Lahue (1976)
   > Bullock (1986)
   > Farley & Alkon (1987)
   > Schreurs (1989)
   > Ricker, Brzorad, & Hirsch (1986)
2. General reviews of the cellular mechanisms of nonassociative and associative learning
   These reviews contain discussion of data obtained from studies of a wide range of invertebrate animals.
   > Carew & Sahley (1986)
   > Davis (1986)
   > Byrne (1987)
   > Farley & Alkon (1985)
3. General reviews of the cellular mechanisms from specific invertebrates
   Many of these reviews have appeared in the book *Encyclopedia of Learning and Memory* (Squire, 1992). If your library does not have a copy, acquire it through interlibrary loan.
   > *Aplysia* Hawkins (1991); Byrne (1992); Castellucci (1992); Carew (1992)
   > Crayfish Krasne (1992)
   > *Drosophila* Tully (1991); Dudai (1992)
   > *Hermissenda* Alkon (1987); Crow (1992)
   > Honeybee Menzel (1992); Menzel, Hammer, Braun, Manelshagen, &
   >   Sugawa (1991)
   > *Limax* Gelperin (1992)
   > *Pleurobranchaea* Gillette (1992)
   > *Tritonia* Frost (1992)
4. Some neglected models of invertebrate learning
   The articles in this section cite some of the more popular simple system models in the 1960s and 1970s. As interest in the use of molluscs has grown it is easy to lose sight that at one time protozoans and other invertebrates were the models of choice for many invertebrate researchers.
   > Protozoan articles of interest:
   > > Applewhite (1972)
   > > Applewhite & Gardner (1971)
   > > Applewhite & Morowitz (1966)
   > > Reynierse & Walsh (1967)

continues

Exhibit 7-1, Continued

Earthworm articles of interest:
    Aranda, Fernandez, Celume, & Luco (1968)
    Blue (1976)
    Roberts (1962)
    Wayner & Zellner (1958)
Cockroach articles of interest:
    Aranda & Luco (1969)
    Brown & Noble (1967)
    Pak & Harris (1975)
    Reep, Eisenstein, & Tweedle (1980)
    Willner (1978)

---

nervous systems can understand. People are not able to see, for example, unless light, striking the retina, produces a movement of molecules; a honeybee will not be able to smell the scent of a rose unless the flower gives off molecules that can be detected by other molecules located in the appropriate membranes of the bee; a crab cannot detect the grip of a predator unless the pressure produces a movement of molecules. There is every reason to believe that the molecular languages of vertebrates and invertebrates have many points of similarity. Recall from the first chapter that the mechanisms involved in, for example, the transmission of graded and action potentials, the operation of ion channels, and the influence of neurotransmitters are similar between invertebrates and vertebrates.

By discovering how molecules affect the activity of the nervous system, scientists will be better able to decipher the "nerve language" associated with learning. To "get at" these molecules and observe "nerve language" in action, invertebrates' nervous systems have been discovered to be excellent listening posts. With the sophisticated tools of the neuroscientist and behavioral geneticist, "wiretaps" can be established that permit researchers to eavesdrop on the physiological and biochemical changes associated with the nerve language of learning.

What Are Some Trends in the Evolution of Nervous Systems?

Before we discuss some of the invertebrate data, it will be helpful to say something about the evolutionary development of the nervous system. Doing so will reinforce for you the idea that the nerve language of certain classes of invertebrates does indeed share properties with the human nerve language.

The simplest animals are the *Protozoa*. Some of the more familiar animals of this group are the paramecium and amoeba. Protozoans are unique in that all the activities necessary for life, such as respiration, reproduction, and feeding are performed within a single cell without the aid of tissues and organs. Yet despite their uniqueness, there are some striking similarities, such as membrane structures, behavior, and the use of metabolic pathways.

The most primitive animals that possess specialized nerve cells arose over 700 million years ago and are known collectively as the *coelenterates*, an example of which includes such organisms as the freshwater hydra and the jellyfish.

One of the more interesting aspects of these animals is the presence of a rudimentary nervous system. Neurons are located regularly over the surface of the animal, and, like a net, they are in contact with those nearest to it. The effect of such an arrangement is that a stimulus applied to any part of the animal will be directed to all parts, much like sticking your finger into a cup of jello will make the whole mass move, or dropping a stone in a bowl of water will produce radial waves. The propagation of a nerve impulse is not transmitted along a linear chain of neurons, as it will be for all animals more advanced than the coelenterates, but radiates from its point of origin. Such a system is not conducive to fine control of motor movements.

The flatworms are the next more complicated group of organisms. Perhaps the most well-known are the planarians, made famous by the work of James McConnell in his studies of learning and memory (see, e.g., McConnell & Shelby, 1970). Of particular importance for those interested in behavior is the fact that the appearance of a brain, bilateral symmetry, polarized neurons, the rudiments of muscle tissue, and definitive anterior and posterior ends—with the anterior end containing a head, "eyes," and other sense organs—first make their appearance in flatworms such as planarians. Moreover, it has been suggested that these and other features of the planarian such as the presence of multipolar neurons, dendritic spines, and neurotransmitter (chemical messengers) substances make these animals ancestors to the human brain (Sarnat & Netsky, 1985).

The nerve cells in planarians and other advanced flatworms are arranged in the shape of a ladder. The two sides of the ladder correspond to what are called nerve cords and constitute the first appearance in evolution of what is called the *central nervous system*. At periodic intervals along the length of the ladder emerge nerves that receive stimuli from, or deliver impulses to, various specific body regions. This, too, represents quite an advance in that it is the first appearance in evolution of what is called the *peripheral nervous system*. In both the central nervous system and the peripheral nervous system, a nerve impulse is propagated along a linear chain of neurons.

The advantage of a central and peripheral nervous system over the arrangement provided by a nerve net is profound. Unlike the jellyfish, the planarian and all other animals more advanced than the planarian need not respond to environmental stimuli with the entire body. Instead, fine motor control is possible because a stimulus will generate a nerve impulse that is carried along the nerve cord to a particular nerve. This nerve will activate a specific organ, or organs, and a response appropriate to the environmental stimulus will occur.

Another interesting feature of the planarian that deserves comment is that, as mentioned previously, they possess bilateral symmetry. If one were to divide such an animal down the middle, the left and right halves would be mirror images of each other—just as in humans. An animal so constructed usually prefers one direction of movement, and fortuitously, it is just this direction in which the head is located—again, just as in humans. The head region contains specialized sense organs. Moreover, the nerve cords and nerve endings become more numerous in this end of the animal. The development of a specialized, permanent head is known as *cephalization*. There are two advantages of devoting a portion of the body to a head region. First, because the head is the leading edge of the animal, it is in a better position to find and capture food. Second, the movement of the animal is best controlled by placing the sensory systems inside the head.

The next evolutionary advance also occurs in a group of worms, in this case the earthworms. Earthworms possess several important advances over flatworms. Like the flatworms, however, they are bilaterally sym- metrical and have a definite anteroposterior orientation. In addition, the tendency for nerve cells to be grouped in the head region is repeated. Among the advances made by earthworms is the development of a diges- tive apparatus consisting of a tubular gut running from a well-developed mouth to an anus. The gut allows the worm to digest food independently of its movement. Another advance—which is perhaps the most distin- guishing characteristic—is the division of the body into similar segments. The segments can respond individually or as a group and can be modified

to perform specialized tasks. The human body contains repeating segments as well. To illustrate this, run your finger along your spinal column.

The next significant advance occurs in a group of animals known as
arthropods. Some well-known members of this group include insects (ants,
bees, etc.), crustaceans (crabs, crayfish, lobsters, etc.), and chelicerates
(spiders, scorpions, etc.). A segmented body plan and a similarity in the
general layout of the nervous system suggest that arthropods arose, directly, or indirectly, from annelids. There are some significant differences,
however. Perhaps the most obvious difference is that arthropods have a
protective and supportive exoskeleton and articulated appendages. Arthropods also have a fixed number of body segments, whereas in most
annelid species, the number of segments are variable. In addition, arthropods have a skeletomuscular system that enables them to be the most
agile of all invertebrates. The zenith of nervous system development in
invertebrates can be found in this group of animals. Although the nervous
system of arthropods is based on the annelid plan, the arthropods have
taken it a step further in the development of, for example, highly specialized sense organs such as the compound eye, antennae bearing chem-
oreceptors, several types of mechanoreceptors, and motor organs. Arthropods are remarkable in having muscle fibers that are controlled by
several types of neurons. These neurons can (a) bring about rapid, brief
contractions, (b) bring about slow but sustained and powerful contractions, or (c) prevent contraction.

In contrast to annelids, the behavior of arthropods is rather impressive. There is an astonishing variety of courtship rituals, social structures, communication systems, and food-gathering strategies, to name
but a few.

The last group of invertebrates that we will look at are the molluscs.
Some well-known examples include snails, clams, and octopuses. After
the arthropods, there are more molluscs than any other invertebrate and
twice as many molluscs as there are vertebrates. The body plan of a
mollusc suggests a segmented ancestor, and consequently, their relationship to annelids has long been postulated. Molluscs are unique in that
they can be considered as two animals: a soft body enclosed within a hard
calcareous shell. The molluscan body consists of a defined head, where
photo-, chemo-, and tacto-receptors are localized; a visceral mass, where
most of the organs lie; and a highly muscular region underneath this
mass known as the *foot*. Another distinctive feature is the *mantle*, which
encloses the greater part, if not all, of the body. The central nervous
system is similar to that of the animals already discussed. It consists of

five pairs of ganglia (cerebral, pleural, pedal, parietal, and visceral) located in the head, the visceral mass, and the foot. As we will see, much of what is known about the neural mechanisms of learning has been gained from work with this group of animals.

What Are the Major Strategies for the Cellular Analysis of Learning?

There are three closely related strategies for the cellular analysis of learning. For our purposes, let us call them the "top-down" approach, the "bottom-up" approach, and the "middle-of-the-road" approach. The major difference between them is where the analysis begins. In the top-down approach, a nonassociative (i.e., habituation or sensitization) or associative learned behavior (i.e., classical, instrumental, or operant) of a freely moving intact animal is selected for analysis. Over a series of many experiments, often taking years, the behavior is progressively "dissected away" (along with the animal) until all that is left of the animal's behavior are the neuronal structures, or analogs, of the original behavior.

The second approach, the bottom-up approach, is to begin with the neuronal structures, or analog, of a behavior, and over a series of many experiments (again, often taking years), work up to the behavior of an intact animal. It has probably occurred to you that it might be possible to reduce some of the time necessary for conducting a cellular analysis of a learned behavior by taking advantage of research directed toward the neuronal analysis of sensory or motor systems. This is precisely what the middle-of-the-road approach attempts to do. Here, one selects a behavior of an invertebrate, such as the escape behavior of the crayfish or of the cockroach, or the olfactory system of a moth, that has been the object of intense investigation at the neuronal level. Having found such a behavior, one attempts to modify it by nonassociative and associative learning procedures.

In contrast to the top-down and bottom-up approaches, the task of uncovering neuronal mechanisms of learning is made somewhat easier with the middle-of-the-road approach because the research program starts with a behavior in which much of the neuronal analysis has already been carried out by other investigators. It is important to keep in mind that most of the research on invertebrates does not involve the study of learning. Rather, it involves how invertebrates process information from their senses, how they move, feed, and reproduce. A large part of this research is directed toward understanding the neural mechanisms involved in these behaviors. When such information is available on a particular invertebrate, it is a relatively easy step to see whether these behaviors can be modified by learning. After all, how an animal uses its senses, moves about, and eats are the necessary elements of any experiment on learning and memory.

Before we move on to an example of the top-down approach, it must be mentioned that whatever your choice of strategies, you will be faced with what are for some of us hard philosophical questions. There are no answers to these questions. I recommend, however, that before embarking on a search for the neural mechanisms of learning, you ask yourself the following three questions:

1. Can an isolated nervous system reveal anything about the behavior of a freely moving organism? You will notice that in each of the three strategies, the researcher dissects, homogenizes, and isolates—and must eventually put the pieces back together again to explain the behavior of an animal.

2. In what sense can one speak of the behavior of a cell, or of a network of cells? In each of the three strategies, the behavior of an intact animal is studied with a variety of stimuli that have texture, appearance, taste, and odor. Depending on such factors as genetic make-up and previous experience, an organism moves toward some stimuli and away from others. In other words, an intact organism exhibits what is commonly referred to as behavior. You must ask yourself if the term *behavior* can be used to explain the actions of cells. If, for example, operant behavior is defined as the goal-directed modification of behavior—such as when a child operates some device to gain access to a room—are we forced to conclude that an increase in the electrical activity of a cell or muscle as a result of the consequence of "its" action is also an example of goal-directed activity? As another example, consider that conditioned stimuli in classical conditioning experiments using an intact organism consist of lights, sounds, odors, tastes, and even the context in which the organism finds itself. Yet there will be a time in your experiments, or reading of experiments performed by others, when these stimuli will be replaced by their neural correlates. In other words, you will stimulate a "visual nerve" instead of turning on a light or stimulate an "auditory nerve" instead of turning on a sound generator. Are such stimulations or analogs equivalent to the "real thing"?

3. Is knowledge of behavioral mechanisms of learning keeping pace with what is known about the neural mechanisms of learning? One of

my goals in writing the apparatus chapter and the chapters on associative and nonassociative learning was to show that not only is existing knowledge about invertebrate learning meager but also that behavioral scientists do not agree on such fundamental issues as what are classical and operant conditioning. (I will have much more to say about this in the next chapter.) If behavioral scientists do not agree on what is classical conditioning, for instance, how can the neural mechanism of such conditioning be uncovered? Before embarking on the search for cellular mechanisms, my advice is to clearly understand what behavioral mechanisms are involved.

In the next few paragraphs, I will illustrate the differences between the three approaches by walking through an example of the top-down approach. As an exercise, keep the aforementioned three questions in mind while going through the example that follows.

As one example of implementing the top-down approach, the behavior of an animal is first observed in its natural habitat so as to obtain clues as to what type of behavior might be trainable. The behavior can be as simple as habituation and sensitization of a withdrawal behavior, or as complex as approaching stimuli associated with food. Often it is not possible or practical to observe an animal under natural conditions. In such cases it is important for you to create an environment that closely approximates that found in nature.

Once a behavior of interest (also known as a *target behavior*) is determined, one begins the process of studying it under laboratory conditions. Let us say, for instance, that after having observed the behavior of crabs under natural conditions, you noticed that the eyes appear to play a significant role in its life. You were particularly struck by the observation that when one crab attacks another, it often attempts to grab the eye of its opponent. How would you study such behavior in the laboratory?

One way is to set up a classical conditioning experiment in which vibration of the shell (called a *carapace*) signals (i.e., vibration becomes a CS) some impending danger to the eye. Although it is difficult, but not impossible, to build a mechanical claw that could grab the eye at the appropriate time, it is more efficient to direct a puff of air to the eye. This air puff will cause the animal to momentarily retract its eye into the carapace in much the same way as the claw of an opponent would. Having found a CS (carapace vibration), a US (air puff), and a UR (retraction of the eye), you have assembled all the elements of a classical conditioning experiment. It remains only to determine whether eye retraction can be elicited by the vibration, that is, whether you can obtain a CR.

Having discovered that the eye of the crab can be classically conditioned, you might decide to move to the next level, at which you analyze the electrical activity of the muscles involved in eye retraction. (You might also wish to stay at this level for a while and answer, as discussed in the previous three chapters, any number of behavioral issues.) Some of the questions you hope to answer might include whether there are any differences in electrical activity between the CS and US and whether the electrical activity to the CS changes over the course of training. You might also decide to use electrical activity as a highly sensitive target response in its own right and do away with the eye all together. One way to do this is to restrict the movement of the eye and record from the muscle associated with eye retraction.

With answers to these and a host of other questions, and ensuring through careful experimentation that changes in electrical activity do indeed reflect changes in behavior, you might become interested in determining what neurons are involved in eye movement and where they are located. This is not an easy task. You will have at your disposal, however, an assortment of dyes, electron and fluorescence microscopes, stimulators, electrodes, and other formidable tools of the neuroscientist to assist in the search. Determining the functional relationships between a behavior and its neural mechanism can be accomplished by, for example, observing the target response when a portion of the nervous system is destroyed, stimulated, transected, or extirpated. Histologically, techniques can also be used in which, for instance, a nerve, is cut and its axon is placed in a dye. The nerve sucks up the dye, and by following the trail with the help of a microscope, it is possible to follow not only its path but also where it makes contact with other nerves. In this way you can begin to uncover a "network" of cells responsible for the target behavior.

After you have identified the relevant neurons and are confident in your ability to repeatedly locate and recognize them (which, by the way, takes considerable experience), you are in a position to progress to the next level. At this point, you create a schematic or circuit diagram describing the network of cells, muscles, and synaptic relationships associated with receipt of the CS and the US. In our example, you must determine what neurons respond to vibration and to air puff. You must also determine what neurons control the UR and CR associated with eye movement. Having accomplished this, the next task is to connect each of the diagrams in such a way that you obtain a complete picture from receipt of the CS to the elicitation of a CR. You can imagine the considerable savings in time and energy by starting with a preparation in which such a schematic diagram exists for at least some of the behavior. Starting with such a diagram is the principal advantage of taking the "middle road." There is, of course, no guarantee that the behavior represented in the diagrams can be trained. New knowledge, after all, is the purpose of discovery.

With diagram in hand, you are now in a position to perform some rather exciting experiments. You might, for example, be interested in how the components of your network change over the course of training. Does the network extinguish when the US is no longer administered? Does it grow with experience? Does it change if you stop presenting the US after each CS presentation?

Experiments such as the one just outlined can be performed by electrically (or chemically) stimulating parts of the network and recording from the appropriate cells. This is most readily done by dissecting away most of the animal and leaving the relevant portions of the nervous system intact. It no longer becomes necessary to use vibration or air puff, because the neural activity associated with these stimuli can be evoked directly by applying a small amount of electrical current or chemical stimulus. Moreover, it is no longer necessary to observe an eye retraction (or any other CR or UR), because you can record directly the neural activity of the cell responsible for such retraction. At this point, you now have what is known as a *neural analog* of classical conditioning. Such analogs can be developed for habituation, sensitization, and instrumental and operant conditioning. A second method for studying the operation of networks is to use your home computer to model the behavior of the network.

It is possible to reduce the preparation further by using some rather sophisticated techniques to actually eavesdrop inside the operation of a cell and determine the molecular events associated with learning. As we will soon see, such reduced preparations can provide remarkable amounts of information about the mechanisms of learning and memory.

In the next two sections, I will present an overview of some of the mechanisms that have been discovered for nonassociative and associative learning. This treatment will be necessarily brief because a complete understanding of these mechanisms requires a working knowledge of physiology, biochemistry, and biophysics. A number of highly readable reviews have appeared over the years and are listed in Exhibit 7-1. I strongly encourage you to read the articles cited in the issue section before embarking on a search for neuronal correlates of behavior.

A Selection of Cellular Models of Nonassociative Learning: Habituation and Sensitization

Nonassociative learning (see p. 37) is a form of behavioral modification that involves the association of one event, such as when the repeated presentation of a stimulus leads to an alteration in the strength or probability of a response. The two principal types of nonassociative learning are habituation and sensitization. Much of our information about the cellular mechanisms underlying nonassociative learning has come from the work of Eric Kandel and his many colleagues studying the large marine sea snail, *Aplysia*.

Aplysia is an excellent example of why invertebrates are useful in the analysis of learning. It has only a few thousand neurons, which are relatively few compared with the billions of neurons in mammals. In addition, many of the neurons are easily located and identified upon examination. This permits the researcher to trace the connections between individual neurons and create "wiring diagrams" for various behaviors. I will present the major features of one of these circuits—the gill-withdrawal reflex—in what I am afraid is a rather torturous fashion to illustrate that it is indeed possible to gather detailed information on "nerve language" and that this nerve language, properly deciphered, is useful in the analysis of learning.

The gill-withdrawal reflex is a defensive reflex in which tactile stimuli applied to the "fleshy spout" (known as the *siphon*) elicits a rapid contraction of the gill. Through much research conducted over many years, the circuit diagram for this reflex is now known.

The fleshy spout contains 24 sensory neurons that respond to tactile stimuli such as touch, water turbulence, and air puff. The rapid withdrawal of the gill is made possible by only six motor neurons. That portion of the sensory and motor neurons containing their respective nuclei meet in a structure containing other such cell bodies, known as the *abdominal ganglion*. In the abdominal ganglion, the axon of each sensory neuron is directly connected to each of the six motor neurons. In addition to these direct connections, there are indirect connections in which the sensory neurons make contact with the motor neurons through a second set of neurons that are designed to integrate information. These neurons are known as *interneurons*.

Behavioral research on habituation in *Aplysia* has demonstrated that if tactile stimuli are applied to the spout, the gill will close, only to open a few moments later. If this stimulus is repeated at regular intervals, the gill will no longer close—it has habituated to touch. By placing micro-electrodes into both the sensory and motor neurons, it was found that such a pattern of tactile stimulation progressively reduced the number of nerve impulses in the motor neurons responsible for gill withdrawal. This effect was caused by an inability of the cell body in the motor neuron to generate a nerve impulse that would be of sufficient strength to trigger a gill withdrawal. Technically speaking, there was reduction in the size of the excitatory postsynaptic potentials (EPSPs) in the motor neurons. Subsequent experiments have shown that the depression of the EPSPs was produced by a decrease in the amount of neurotransmitter released by each nerve impulse from the sensory neurons. There was no change in the ability of the motor neuron to trigger a gill withdrawal. What changed was the amount of transmitter available to stimulate the motor neurons.

Additional experiments were designed by Kandel and his associates that determined the chemical mechanisms responsible for the depressed transmitter release. When a neuron generates an action potential, there is an influx of calcium ions into the axon terminals, and this in turn stimulates the release of neurotransmitter into the space between one neuron and the next. It was found that habituation produced a diminished influx of calcium ions into the axon terminals of the sensory neurons. Therefore, the cellular basis of habituation in *Aplysia* was found to be a decrease in calcium current, thereby producing a decrease in the amount of neurotransmitter released into the synapse. This in turn decreases the excitation of the motor neuron, producing a weakened gill withdrawal reflex.

The neural mechanisms for sensitization in *Aplysia* are in many ways the opposite of what is found in habituation. Recall that sensitization exhibits the augmentation of a response to the presentation of a stimulus. Sensitization of the gill withdrawal response can easily be demonstrated by electrically shocking, or pinching, the tail of the animal prior to ha-bituation trials. The result of such a treatment is a gill withdrawal reflex that is stronger than normal. Microelectrodes placed in the motor neurons indicate that the EPSPs are larger, and more nerve impulses are produced. In addition, an analysis of the chemical mechanisms producing sensitization reveals that there is a greater influx of calcium ions in the terminals of the sensory neurons. This in turn produces an increase in the amount of neurotransmitter released, which, of course, stimulates the motor neuron.

It has also been discovered that neurons known as *facilitory interneurons*, which are connected to the sensory neurons, help to increase the amount of neurotransmitter during the presentation of a sensitizing stimulus. If you are interested in obtaining more information about the cellular mechanisms of habituation and sensitization in *Aplysia* and other invertebrates, consult the references in Exhibit 7-1.

A Selection of Cellular Models of Associative Learning: Classical and Operant Conditioning

Associative learning (see p. 38) is a form of behavioral modification that involves the association of two or more events, such as the association between two stimuli or between a stimulus and a response. As I have already discussed, the three principal types of associative learning are classical conditioning, instrumental conditioning, and operant conditioning. As is the case with sensitization and habituation, much of our information about the cellular mechanisms underlying associative learning has come from work with *Aplysia*. In this section, I will outline some of these mechanisms. I will also discuss some of the operant conditioning models developed for *Aplysia* and for an insect—the locust.

Classical Conditioning in *Aplysia*

The gill withdrawal reflex, which we have already examined in some detail, can also be classically conditioned. The CS is a touch to the spout, and the US is an electric shock. (The issue of whether touch should be considered a CS will be discussed in the next chapter.) Kandel and his associates discovered that the cellular mechanisms for classical conditioning are similar to those we have looked at for sensitization. Such a finding suggests that nonassociative learning phenomena form the building blocks of more complex learning. Classical conditioning in *Aplysia* is produced by an increase in the influx of calcium ions in the terminals of the sensory neurons. This in turn produces an increase in the amount of neurotransmitter released which, like sensitization, stimulates the motor neuron for gill withdrawal. The principal difference between sensitization and classical conditioning in *Aplysia* is that, for classical conditioning to occur, the sensory neurons must be active prior to the activation of the facilitory interneurons by the US.

The view that classical conditioning involves changes in presynaptic neurons is not universally accepted. A series of studies by Daniel Alkon and his colleagues on another mollusc—*Hermissenda*—suggests that learning does not take place at the synapses. By pairing light—to which *Hermissenda* is normally attracted—with rotation, it is possible to condition a reduced tendency to move toward the light. Using the same techniques employed by Kandel, Alkon was able to show that the important changes that occur during learning in his mollusc took place in the neuronal membranes of the photoreceptor cells and not at the synapses.

The results obtained with *Hermissenda* and with *Aplysia* suggest two types of cellular mechanisms. Research is currently being conducted with these and other invertebrates such as the garden slug *Limax*, honeybee, fruit fly, and another marine gastropod, *Pleurobrachaea*, to determine how general and correct these two mechanisms are. A review of the rather complex issues involved can be found in Exhibit 7-1. An enjoyable account of Alkon's search for the cellular mechanisms involved in his learning research with *Hermissenda* can be found in his book, *Memory Traces in the Brain* (Alkon, 1987).

Operant Conditioning in Invertebrates

In contrast to habituation, sensitization, and classical conditioning, there are few attempts to develop cellular models of operant behavior. The principal difficulty is finding a behavior that is controlled by its consequences yet is amenable to cellular analysis. Attempts to create such models are available for position learning in crabs, roaches, and *Aplysia*. The only model, however, in which some progress has been made at determining the cellular mechanisms is leg position learning in the locust (Forman & Zill, 1984) and head waving in *Aplysia* (Cook & Carew, 1989a, 1989b, 1989c).

You may recall from chapter 3 that Robin Forman developed a training situation for locust in which the animal is trained to maneuver a leg into an arbitrarily selected position. When such a response is made, a reinforcement is delivered. Depending on the experiment, reinforcement might consist of turning off a heat lamp directed to the animal's leg, or in the case of positive reinforcement, access to grass.

The circuit diagram for locust leg movement had been worked out, and this enabled the researchers to record leg movement from the identified motor neuron responsible for the adjustment of leg position. Through a long series of experiments, Tosney and Hoyle (1977) have shown that it is possible to manipulate the electrical activity of this neuron by delivering reinforcement when the firing rate of this neuron reached an arbitrarily selected threshold. The researchers suggested that the neural mechanisms of learning are reflected in changes in pacemaker activity within the motor neuron rather than changes in synaptic input to the motor neuron. Exhibit 7-1 includes references to articles on other models of operant conditioning. In comparison with nonassociative and classical conditioning, there remains much work to be done.

Summary

In this chapter we examined some of the factors that make invertebrates well suited for a cellular analysis of learning and looked at some of the trends in the evolution of nervous systems. Three of the major strategies used to develop cellular models were outlined, and an illustration was provided for the top-down approach. The final portion of the chapter was devoted to a summary of some of the cellular mechanisms that have been discovered for nonassociative and associative learning.

Discussion Questions

- Why are invertebrates useful for a cellular analysis of learning?
- What advances are present in the planarian that foreshadow those found in humans?
- What are three major strategies associated with the analysis of invertebrate learning?
- Which of the three strategies do you favor?
- Can an isolated nervous system reveal anything about the behavior of a freely moving organism?
- In what sense can one speak of the behavior of a cell or of a network of cells?
- Is knowledge of behavioral mechanisms of learning keeping pace with what is known about the neural mechanisms of learning?

- How would you conduct a search for the cellular analysis of learning?
- What are the cellular mechanisms for habituation in *Aplysia*?
- Consult your library and discuss the similarities and differences in the cellular mechanisms for classical conditioning in three invertebrates.

8 Issues in Invertebrate Learning

Preview Questions

- What are some of the procedural issues in invertebrate learning?
- Can we develop categories of behavior?
- Are there inconsistencies in the definition of learning phenomena?
- Should all behavior modified by its consequences be considered operant conditioning?
- Is there a difference between signaled avoidance and punishment?
- Is there value in reporting the data of individual invertebrates in addition to reporting data based on a group of invertebrates?
- What are some conceptual issues in the interpretation of results?
- Are there differences in the learning ability of invertebrates and vertebrates?
- Is it appropriate to explain the learning of invertebrates in terms of cognitive concepts?
- Can you make a significant contribution to the invertebrate learning literature?

In previous chapters we have: (a) discussed some of the strategies used to study the learning of invertebrates, (b) identified the major behavioral techniques available to study invertebrate learning, and (c) surveyed much of the research in nonassociative and associative learning. It is now time to close this primer with a discussion of five behavioral issues that I believe should be considered in the analysis of invertebrate learning.

Discussion of these issues was intentionally delayed until you obtained the necessary background information. Some of these issues have been briefly touched upon in earlier chapters, and others you may have recognized in the course of reading this book. In fairness to you, I must mention that there is no consensus in the invertebrate learning community as to the importance of these issues. Your instructors may agree with some of these issues and disagree with others, and, of course, you may have your own opinions. The goal of this chapter is to encourage you to think about these issues and to develop your own opinion, preferably before embarking on a program of invertebrate learning research.

These five issues will be divided into two broad categories: (a) **Methods,** or, procedural issues in the measurement of behavior, and (b) **mechanisms,** or, conceptual issues in the interpretation of results.

Under the category of methods, we will examine the need to:

- Develop taxonomies of invertebrate learning paradigms
- Clarify inconsistencies in the definitions of learning phenomena
- Encourage the reporting of individual-level data

Under the category of mechanisms, we will discuss the need to:

- Determine the extent of phylogenetic differences in vertebrate/invertebrate learning phenomena
- Evaluate the suitability of using cognitive explanations of invertebrate learning phenomena

Those of my readers who have some acquaintance with the invertebrate learning literature conducted during the 1950s and 1960s will recognize many of the issues listed here from journal articles and from numerous workshops and informal discussions of invertebrate learning (Bullock & Quarton, 1966; Teyler, Baum, & Patterson, 1975). You who are new to invertebrate research will find it instructive to read some of the articles written during that time period. Many of these workshops and articles are cited in Appendix A. By reading some of those long-forgotten articles, you will experience personally one of the ironies of invertebrate learning research—that despite an increase in fundamental knowledge, many of the issues listed earlier are not resolved and are simply passed along from one generation of behavioral scientists to the next.

For example, it is somewhat disconcerting that 40 years after the "golden age of learning theory," Amsel (1989) must remind us to reject the trend to characterize invertebrate behavior as cognitive; Staddon and

Bueno (1991) counsel that an understanding of behavior must precede the understanding of brain–behavior relationships; Bitterman (1975) reminds us about the need to consider the role of evolution in learning research; Hirsch and Holliday (1988) warn us about the need to report individual data; Gormezano (1984) reminds us that what are thought to be classical conditioning procedures may not be classical at all; and Schreurs (1989) takes this a step further and suggests that classical conditioning in *Aplysia*—perhaps the most important invertebrate animal used in the simple systems strategy—may not be classical conditioning at all!

What Are Some Procedural Issues in the Measurement of Behavior?

Can We Develop Taxonomies of Learning Paradigms?

As you begin to read the invertebrate learning literature, you will discover that there is a trend to neatly characterize associative *learning* into either classical or operant conditioning. This tendency is present in my own work (Abramson & Feinman, 1990b) and in several recent reviews of invertebrate learning (e.g., Byrne, 1987; Carew & Sahley, 1986; Farley & Alkon 1985; Hawkins, 1991).

Invertebrate learning *paradigms* are not so neatly characterized! Recall from the chapter on classical conditioning (chapter 5) and from our discussion of the methods used to study invertebrate learning (chapter 4) that there are at least eight different ways to study classical conditioning of invertebrates including conditioning of (a) proboscis extension in insects, (b) eye withdrawal in crustaceans, (c) taste aversion in snails and molluscs, (d) contraction in worms, (e) suppression of phototaxic behavior in *Hermissenda*, (e) contraction of gill withdrawal in *Aplysia*, (f) olfactory conditioning in *Drosophila*, and (g) conditioning of "emotional responses" in *Aplysia*. I am sure it has occurred to you, as it has to me, that before it can be concluded that each of these procedures measure the same "thing"—that is, classical conditioning—we must know the relationships between these procedures and how they are influenced by parametric manipulations of variables such as stimulus intensity, interstimulus interval, and partial reinforcement, which have seldom been investigated. It may be helpful to think of this issue in terms of baking a cake. A cookbook, for example, may contain several different recipes for chocolate cake. Although each recipe ensures that you are making a chocolate cake, there will be significant differences in taste, texture, and style, depending upon the ingredients and recipe. Some cakes, for example, are light as a feather, others are as dry as sawdust; some are "no-stick," and others are microwavable. How significant these differences are can be determined only by comparing the cakes.

Similarly, questions about the similarities and differences between these very different experimental procedures can be answered most effectively by creating a classification of behavioral measurement in which classes of behavioral observations, under varying testing conditions, are arranged in a meaningful and consistent way. The advantages of creating behavioral taxonomies are readily apparent from the progress made in the animal orientation literature (e.g., Bell & Tobin, 1982; Fraenkel & Gunn, 1961; Schöne, 1984; Van der Steen & Maat, 1979).

In an excellent article on the importance of classification in studies of learning and memory, Tulving (1985) describes six ways in which a classification scheme can advance the field of learning and memory. First, it would lead to a comprehensive understanding of learning and memory. Behavioral scientists would now speak the same language, and a taxonomy would give some theoretical structure to the design and analysis of experiments. Second, a taxonomy would replace the use of general categories such as "classical conditioning" and "operant conditioning" with detailed descriptors of the procedures. This would have the immediate effect of limiting the generalizations of empirical facts across species. Third, theoretical processes could be specified with greater precision because they would be anchored in a strong classification system. Fourth, novel procedures and results could be described easily in terms of the amount of deviation from specified categories. Fifth, a classification system would stimulate a comparative approach to the study of learning and memory. Sixth, it would get scientists interested in behavioral problems articulated in the same "language."

The importance of creating such categories and the controversies they engender has long been recognized in the study of vertebrate learning. Skinner (1935), for instance, proposed two types of learning: Type I and Type II. Hilgard and Marquis (1940) also proposed two types of learning: "classical conditioning" and "instrumental conditioning," as did Maier and Schneirla (1942), who maintained a distinction between classical conditioning and selective, or "trial-and-error" learning. Spence (1956), however, believed that there are three categories of learning, adding "selective learning" to those suggested by Hilgard and Marquis. Amsel (1972) also postulated three categories, adding "pure" classical to Pavlovian conditioning and instrumental conditioning. Hearst (1988) also divided learning into three types, and it should be noted that there are also those who believe in only one type of learning. This position is represented, for instance, by the one-factor theories of Thorndike (1911), Pavlov (1927), Guthrie (1935), Hull (1943), and Bindra (1976). In contrast to the one-factor theorists is the work of Gagné (1965), who postulated eight types of learning; Tolman (1949), who postulated six types; Woodworth (1958), who postulated five types; and Razran (1971), who postulated four. Other attempts at behavioral taxonomies can be found in the work of Bitterman (1962) and Dyal and Corning (1973). The work of Dyal and Corning is especially relevant because it deals exclusively with invertebrate learning.

When reviewing the invertebrate learning literature, it is important to keep in mind that there is no agreement in the vertebrate literature as to how many categories of learning there actually are. Clearly, this disagreement within the area of vertebrate learning must serve as a warning against the use of general categories in the study of invertebrate learning. As Bitterman (1962) noted over 30 years ago, "Classification is not merely a matter of taste."

Let us consider for a moment how the lack of an accepted behavioral category presents problems for the classical conditioning of *Aplysia*. Recall from our discussion of classical conditioning that a distinction is often made between alpha conditioning and classical conditioning. *Alpha conditioning* refers to a response that is elicited by the CS prior to training. The problem occurs when the response elicited by the CS resembles the response one uses as the index of learning. Some members of the invertebrate learning community have forcefully argued that this is an artificial distinction, for at the cellular level, alpha responses and "true" classical responses may not represent separate processes but instead represent separate points on a response threshold continuum (Carew, Abrams, Hawkins, & Kandel, 1984). However, other members of the invertebrate learning research community, myself included, believe that the alpha response is not the same as a classical conditioned response, although both may obey the same laws of conditioning. Therefore, in the minds of some, the question of classical conditioning in *Aplysia* remains open (Farley & Alkon, 1985, 1987; Schreurs, 1989). How is one to decide which conceptualization is the correct one? The answer is that one must place a greater emphasis on the analysis of invertebrate behavior.

This *Aplysia* picture can be complicated further by making the plausible assumption that a CR to electric shock is formed because the withdrawal response (the CR in such experiments) reduces the amount of electric current flowing through the animal. Rats placed in similar conditions often are observed to "fold up" to stimuli predicting shock. Such behavior on the part of the rat reduces the amount of current flowing through the animal, and therefore, reduces the amount of physical discomfort. If this assumption is confirmed in *Aplysia*, by measuring the amount of current flowing through the animal during a withdrawal CR, then alpha conditioning in *Aplysia* would formally fit the definition of instrumental training in that a preexisting response is reinforced! In our example, the reinforcement would be shock reduction, and the preexisting response would be the slight withdrawal response elicited by the CS prior to any training with the US. The *Aplysia* physiological and biochemical models that we discussed in chapter 7 could, therefore, just as easily be applied to instrumental conditioning as to classical conditioning! How is one to decide which behavioral model to use? The answer once again is that one must place a greater emphasis on the analysis of invertebrate *behavior*.

Presenting a category of learning is beyond the scope of this book. However, I suggest that in your own research you examine the several categories that are now in existence and when reporting the results of your research, link your procedure to one (or several) of the categories. By following this suggestion, your procedure can easily be identified and related to others. A good starting point might be the categories proposed by Dyal and Corning (1973) for classical, instrumental, and operant conditioning, that proposed for classical conditioning by Gormezano et al. (1983), which was discussed in chapter 4, or that proposed for instrumental and operant conditioning by Woods (1974). Woods' (1974) classification of instrumental conditioning identifies 16 categories of conditioning based on the presence or absence of a discriminative stimulus and the desirability of the reward!

Are There Inconsistencies in the Definitions of Learning Phenomena?

A second procedural issue in invertebrate learning, which follows logically from the first, is the lack of consensus in categorizing many invertebrate conditioning techniques. This problem is a recurring one in the invertebrate learning literature and has been addressed in at least two workshops (Bullock & Quarton, 1966; Teyler et al., 1975). If the study of invertebrate learning is to flourish, categorical issues must, in my opinion, be addressed once again. To illustrate some of the inconsistencies, I will select examples from operant conditioning, signaled avoidance conditioning, and punishment.

Operant Conditioning

In contrast to studies of classical conditioning and of habituation and sensitization, there has been limited application of the techniques of operant conditioning to invertebrates. As originally conceived, operant behavior is characterized by the "goal-directed" motor manipulation of the environment, or as Hoyle (1988) has stated, the "reinforcement of endogenously generated movements" (p. 435). In place of the goal-directed modification of behavior that was the hallmark of the Skinnerian system, operant conditioning of invertebrates generally consists of *any* behavior sensitive to response–reinforcer contingencies. Thus, operant conditioning of invertebrates is now deemed to include such procedures as manipulating body position or a single appendage in response to aversive stimulation (i.e., the Horridge procedure; see chapter 3); running against a taxes or kinetic preference, as, for example, when roaches (which as a rule, normally run away from lighted places) are shocked for entering a dark compartment; and learning in various mazes and runways. These procedures, however, might not constitute operant behavior.

A requirement of operant conditioning has been that species-typical behavior is minimized by interjecting a "novel" behavior, such as a lever press, or a nonarbitrary response brought under the control of a *discriminative stimulus* or cue placed between the animal and the animal's receipt of some reward. In this way, the experimenter demonstrates that the animal has learned not only how to operate some device but also how to use it. An example of the latter would be training an animal to run around a barrier to obtain food on cue. In vertebrate learning experiments, animals are trained to *operate* a manipulandum to obtain a positive event or remove a negative one or to run down an alley at a speed selected by the experimenter, or they are given a series of choices. In the best cases, "novel" behavior is created or "shaped." Because novel behavior is presumed to be created, operant behavior (in contrast to most instrumental runway or maze behavior) is generally thought to be more complex than classical conditioning. Before you engage in a study of operant conditioning in an invertebrate, I would urge you to consult Schick (1971), Catania (1973), Lee (1988), and Staddon and Zhang (1989) for discussions of what constitutes operant behavior.

In addition to differences of opinion as to what does and does not constitute an operant response, there are also differences of opinion concerning how it should be measured. As we have seen, some invertebrate operant techniques involve the modification of innate responses. Other techniques thought to tap into the operant behavior of invertebrates are the instrumental maze and runway procedures. As the vertebrate data suggest, it is often difficult to separate classical conditioning effects from operant behavior in these situations. For instance, maze experiments can easily be interpreted in terms of classically conditioned approach responses to stimuli preceding reinforcement. The same can be said of the popular free-flying honeybee technique, in which bees are trained to fly from the hive to the experimental arena. A bee, for example, may not fly to the target because it "expects" food but rather because the target has been associated with food in much the same way we associate a lemon with sour taste. One might be more confident that an invertebrate technique is measuring an operant if it can be shown that (a) the operant response minimizes species-typical behavior, (b) some property of the response such as its rate or force can be trained, (c) the response no longer occurs when such responses postpone the delivery of reward, and (d) the response can be brought under the control of a cue. In my opinion, the problem of what constitutes operant behavior and its measurement must be solved before a meaningful comparative or physiological analysis can be undertaken in invertebrates.

The number of invertebrate operant conditioning experiments would become very small indeed if the response-contingent modification of innate responses and the maze and runway experiments were no longer considered examples of operant conditioning. One of the few examples of an invertebrate operant conditioning situation that involves the shaping of an innate response is the work of Forman (1984; Forman & Zill, 1984), who trained locusts to receive a positive event or remove a negative one by maneuvering a leg to an arbitrarily selected position. Studies in which species-typical behavior is minimized by training an invertebrate to operate a manipulandum, although rare, are not as rare as one might think. Examples of these devices were presented in chapter 3.

The first use of a manipulative response with an invertebrate was Rubadeau and Conrad's (1963) demonstration of food-reinforced lever-press behavior of the roach! It is amazing, considering today's standards of reporting research, however, that this article contained absolutely no data. A lever-press device has been used for the solitary bee *Meloipona anthidiodes* (Pessotti, 1972). A preliminary report describing lever-press performance for crayfish has also appeared (Olson & Strandberg, 1979). Lever press has been demonstrated in work with *Aplysia* (Downey & Jahan-

Parwar, 1972) and, most recently, an unusual study of bar press with *Helix* has been described in which animals work for neuronal stimulation (Balaban & Chase, 1989). Each of these devices is illustrated in chapter 3.

In none of these examples of operant lever pressing has the analysis been carried very far. The lack of behavioral data in these operant paradigms is surprising. It is not known, for example, whether any of the invertebrates can maintain responding on a simple schedule in which every second response is rewarded (FR-2). Moreover, there are no data on the stability of the lever-press response over time, whether some property of the response such as its rate or force can be modified by contingencies of reinforcement, whether the response is sensitive to omission contingencies, or whether the animal can learn to respond on cue. Clearly much work remains to be done.

Perhaps the most detailed behavioral analysis of operant lever-press conditioning of an invertebrate is found in the green crab. Crabs are rewarded for each lever press with a bit of squid. It has been demonstrated that each lever press does not need to be reinforced and that the animals can learn to press one of two levers (Abramson & Feinman, 1990b). In addition, it has been reported (Feinman, Korthals-Altes, et al., 1990) that at least some crabs can perform when the ratio of lever presses to reinforcement is nine (i.e., FR-9). In other words, some crabs are capable of pressing a lever nine times in order to gain access to a bit of food. Other research (Abramson, unpublished observations) suggests that green crabs will continue to press when food is delivered after a FI of time has elapsed (i.e., FI schedule; see p. 166). However, as is the case with every invertebrate lever-press demonstration, it is not known whether the crustacean lever-press response is sensitive to omission contingencies and whether some property of the lever-press response is sensitive to reinforcement. In other words, we do not know if the crab will continue to press the lever when such pressing prevents food, and we do not know if it is possible to train the crab to press the lever with an arbitrarily selected force. Until such data are available, the discovery of operant behavior in the crustacean must be considered tentative.

Signaled Avoidance and Punishment

In addition to problems in defining what constitutes operant conditioning of invertebrates, a second area in which confusion exists is avoidance conditioning. Avoidance techniques have been used in the vertebrate literature both as a diagnostic tool to tease apart classical conditioning from operant conditioning and, more recently, to test cognitive versus noncognitive interpretations of learning.

There have been numerous studies in the invertebrate literature purporting to demonstrate avoidance. Many of these studies contain the pervasive conceptual error of failing to distinguish between the paradigm of punishment and the paradigm of avoidance. In punishment training, emitting a target response results in the *presentation* of the aversive event. However, in avoidance training, emitting the target response results in the *postponement* or *omission* of the aversive event.

I believe that the failure to recognize this distinction stems from a misunderstanding of an early theory of punishment that stressed the acquisition of competing responses (e.g., Dinsmore, 1954). The passive avoidance theory suggests that punishment is effective because stimuli associated with aversive reinforcers elicit behavior that competes with the target response. An animal engaged in such behavior "avoids" the punishment contingency by withholding the target response. From this perspective, cockroaches (Pritchatt, 1968, 1970) and fruit flies (Booker & Quinn, 1981) reportedly "avoid" shock in the Horridge leg-position preparation, cockroaches "avoid" darkness when light is present (Minami & Dallenbach, 1946; Szymanski, 1912), and ants "avoid" an area associated with X rays (Martinsen & Kimeldorf, 1972). In each of these examples, a response results in the *presentation* of an aversive event. After a number of such response–reinforcer pairings, the probability of response—whether it be the manipulation of an appendage or a more molar behavior— *declines*.

The difference between avoidance and punishment continues to exist when signals are introduced into the punishment paradigm as controls for nonassociative effects. Discrimination is but one of a number of traditional controls for the effect of stimulation per se in conditioning experiments. A widely used technique for flies, intended primarily to provide a target-behavior for genetic analysis (e.g., see Dudai, 1977; Hewitt, Fulker, & Hewitt, 1983; Quinn et al., 1974), is considered by its advocates to represent signaled avoidance. In this mass conditioning situation, which we discussed in chapter 3, fruit flies are trained with two signals, one of which is associated with an aversive stimulus. Because presentation of the aversive stimulus is *presented only when the flies enter* a particular portion of the apparatus, this experiment is properly interpreted as punishment.

This analysis also holds for the avoidance experiments of Mpitsos and Davis (1973) and other experiments in which invertebrates "avoid" the taste of some food. These authors found that animals will stop responding to a food CS that has been paired with an aversive stimulus. Based on our discussion, you should be wondering: (a) Why is this considered by some experimenters as a case of avoidance conditioning rather than a case in which the animal is receiving punishment for eating a distinctive food (e.g., see Carew & Sahley, 1986)? and (b) In what sense can food be classified as a CS? Recall from our discussion of classical conditioning in chapter 5 and our discussion earlier in this chapter that a CS must be neutral. If you agree that a CS should be neutral, then taste does not qualify as a CS. Taste learning may be better qualified as an example in which the association between two USs is learned.

A similar concern is the Conditioned Emotional Response (CER) paradigm of Walters, Carew, and Kandel (1979) developed to measure classical defense conditioning in *Aplysia*. Once again, food is used as a CS and when paired with shock produces withdrawal from food. It must be pointed out that when the CER procedure is used with vertebrates, a lever-press apparatus is used and the CS consists of lights and tones, not food or taste.

This seemingly insignificant change from an arbitrary learned response to one consisting of an innate response may have important practical and theoretical consequences. The effectiveness of punishment, for example, is in fact partly determined by whether the punished response is innate or learned (Boe & Church, 1968).

In contrast to punishment experiments, the number of signaled avoidance experiments that resemble the techniques used with vertebrates is quite small. Signaled avoidance—in which the presentation of the CS leads to an *increase in the probability* of a target response—has been demonstrated in ants (Abramson, unpublished observations), honeybees (Abramson, 1986; Smith et al., 1991) and green crabs (Abramson et al., 1988; Feinman, Llinas, Abramson, & Forman, 1990). It has been less convincingly demonstrated—because of lack of control procedures—in roaches (Chen et al., 1970), earthworms (Ray, 1968), planarians (Ragland & Ragland, 1965), and crayfish (Taylor, 1971). The roach and planarian experiments have the additional complication that the avoidance signals were not neutral. Shock was used as a CS in the roach experiment, and light was used in the planarian work. Negative results have been obtained with houseflies (Leeming, 1985), roaches (Pritchatt, 1970), horseshoe crabs (Makous, 1969), and earthworms (Kirk & Thompson, 1967). Unsignaled avoidance has been demonstrated in bees (Abramson, 1986) and roaches (Longo, 1964). In contrast to vertebrate studies, the invertebrate unsig-

Table 8-1

Representative Sample of Individual Data

Paradigm	Animal	Citation
Habituation/sensitization	Crab	Rakitin et al. (1991)
	Crab	Walker (1972)
	Crayfish	Hawkins & Bruner (1981)
	Cricket	May & Hoy (1991)
	Leech	Debski & Friesen (1985)
Classical conditioning	*Aplysia*	Lukowiak & Sahley (1981)
	Crab	Abramson & Feinman (1988)
	Earthworm	Herz et al. (1964)
	Fly	Nelson (1971)
	Planarian	Thompson & McConnell (1955)
Taste aversion	Crab	Wight et al. (1990)
Instrumental conditioning	Ant	DeCarlo & Abramson (1989)
	Ant	Vowles (1965)
	Bee	Smith et al. (1991)
	Crab	Hoyle (1976)
	Crab	Dunn & Barnes (1981a, 1981b)
	Crab	Abramson & Feinman (1987)
	Crayfish	France (1985)
	Drosophila	Booker & Quinn (1981)
Signaled avoidance	Crab	Abramson et al. (1988)
	Roach	Chen et al. (1970)
Operant conditioning	*Aplysia*	Downey & Jahan-Parwar (1972)
	Bee	Sigurdson (1981a)
	Bee	Grossmann (1973)
	Bee	Pessotti (1972)
	Crab	Abramson & Feinman (1990a)
	Snail	Balaban & Chase (1989)

naled data do not reveal the formation of temporal regularities so often characteristic of vertebrate unsignaled avoidance.

Is There Value in Reporting the Data of Individual Subjects?

In contrast to vertebrate work, there are few reports of individual-level data associated with invertebrate learning experiments. In some cases, as in the honeybee classical conditioning data, the data are often presented as the percentage of animals responding on each trial. In other cases, such as *Aplysia* classical conditioning, the data are presented as group means. Grouped data give little impression of the shape of individual learning curves and do not provide much information about the variation between animals. Moreover, it is often not reported how many animals (if any) are discarded from a test population. Without such data it is difficult to know how many animals from a given population do indeed learn. A similar concern was applied by Hirsch and Holliday (1988) to the fruit fly mass conditioning situation developed by Quinn (Quinn et al., 1974).

Few individual data are published for the classical conditioning experiments involving *Limax*, *Aplysia*, *Apis*, and *Hermissenda*. For example, Greenberg, Castelucci, Bayley, and Schwartz (1987) mention in the caption of their Figure 1 the high degree of variability encountered in behavioral experiments with *Aplysia*. To my knowledge, this is the only reference to individual variability in the *Aplysia* model. As another example, consider some individual data that have recently been reported for the proboscis extension paradigm in honeybees (Smith et al., 1991). Here the data reveal that the group learning curve is not always representative of individual performance. This has been reported also for proboscis conditioning in blow flies (Nelson, 1971).

The lack of individual data is also obvious in the operant conditioning of invertebrates. Individual data are available only for the crab (Abramson & Feinman, 1990b) and bee (Grossmann, 1973; Sigurdson, 1981a). More often than not, when individual operant data are available, it is in the form of chart recordings that suggest long pauses between responses rather than the smooth, steady rate of responding associated with vertebrate operant conditioning (e.g., see the charts of Balaban & Chase, 1989)! I would suggest to you that, at least in the initial demonstration of operant conditioning, your data be presented in the form of cumulative records. This would obviously facilitate comparisons between vertebrates and invertebrates and give some idea of the individual variability of behavior. Table 8-1 presents a representative sample of experiments that present individual data.

What Are Some Conceptual Issues in the Interpretation of Results?

Are There Differences in the Learning Ability of Vertebrates and Invertebrates?

In a series of papers beginning in the 1950s, a number of psychologists have questioned the lack of an evolutionary perspective in animal learning (e.g., Beach, 1950; Schneirla, 1949). Perhaps some of the best known critiques are the work of Bitterman (1960, 1975, 1988) and his associates, in which he demonstrates differences in the learning ability of several species of vertebrate and invertebrate, and the work of Lejeune and her associates on the demonstrated species differences in the ability to discriminate time (Lejeune & Richelle, 1990; Lejeune & Wearden, 1991; Richelle & Lejeune, 1980). (For a recent discussion of phylic differences in learning, see Macphail, 1987.) A reading of the invertebrate learning literature suggests there is a tendency to *minimize* differences in learning ability between invertebrates and vertebrates. For example, Sahley (1984) states that "it is clear that striking commonalities exist between learning in invertebrates and learning in vertebrates" (p. 190). In my opinion, such a statement is premature until more behavioral data are obtained. As we saw in the previous chapters, basic questions remain as to whether, for example, many invertebrates actually exhibit classical conditioning and whether operant conditioning and avoidance are as widespread as many believe.

Arguably, one of the best representatives of the evolutionary point of view is the work of Gregory Razran (1971). Razran's *Mind in Evolution: An East-West Synthesis of Learned Behavior and Cognition* reviews over 1,500 papers on learning and memory. On the basis of this review, he was able to identify four "superlevels" of learning: (a) reactive, (b) associative, (c) integrative, and (d) symbolic. Consistent with the work of other comparative psychologists such as Schneirla (Maier & Schneirla, 1964) and Bitterman (1975), later levels emerge from earlier ones, dominate ancestral levels, and can coexist with lower levels in phylogenetically derivative species. It is not my intention to discuss Razran's position in detail, but it is worth noting that, to Razran, the most *"primitive"* example of associative conditioning is "inhibitory conditioning"—what is commonly known in the vertebrate literature as *punishment* (and in the invertebrate literature as *avoidance*). Moreover, alpha conditioning is considered not as an example of classical conditioning but rather as the evolutionary precursor to instrumental conditioning. Operant conditioning is considered to be the highest form of learning in the associative superlevel.

As an example of how Razran's concepts conflict with the current view that many examples of invertebrate associative conditioning are analogous with vertebrate conditioning, I will discuss two examples—operant conditioning and avoidance.

As mentioned in chapter 6, the phyletic generality of operant conditioning is much less clear than that of habituation and sensitization. The problem of phyletic generality of operant behavior may reduce to how one defines operant behavior. When operant behavior is broadly defined as *any* behavior that is "controlled by its consequences," then operant effects are present in all animal groups. However, if operant behavior is defined in terms of its functional influence on the environment, then the phyletic generality is limited to vertebrates and perhaps a few species of molluscs, crustaceans, and insects. If one accepts Razran's levels position, then most of what passes in the literature as invertebrate operant conditioning models actually confounds a simpler associative level, such as that revealed by aversive inhibitory conditioning (i.e., punishment of innate responses), with operant behavior—the highest level of performance associated with the Associative sublevel.

A similar confounding of primitive and advanced levels appears in signaled avoidance conditioning. If punishment experiments are removed from the category of signaled avoidance, as might be appropriate, a pattern begins to emerge. First, it appears that signaled avoidance conditioning is not a general property of invertebrate behavior. Primitive invertebrates such as houseflies, roaches, and horseshoe crabs do not show such learning, and the ability in planarians, earthworms, and crayfish is suspect.

Secondly, of those invertebrates in which signaled avoidance has been demonstrated, the relevant behavioral mechanism appears to be a straightforward application of simple Pavlovian principles. In an experiment designed to vary the CS termination and US avoidance contingencies, carpenter ants were trained in a shuttle box with vibration as the CS and peppermint odor as the US. Animals learned to shuttle whether or not the response avoids odor (Abramson, unpublished observation). In another study, honeybees were trained to fly back and forth between the hive and the sill of an open laboratory window to take sucrose solution from a target. The target was constructed so that the animal could be shocked when the proboscis was in contact with the solution, and bees learned to fly off the target when stimuli signaling shock were presented. The results from an unpaired control group, a differential conditioning group, and the cyclic nature of the data in the paired group (which resembled a series of Pavlovian acquisition and extinction sessions) suggest that the results can be interpreted in Pavlovian terms (Abramson, 1986). A similar pattern of results is observed in another avoidance experiment using crabs as subjects (Abramson et al., 1988). In contrast, behavioral mechanisms developed to account for signaled avoidance in vertebrates is much more complex.

Is It Appropriate to Explain the Learning of Invertebrates in Terms of Cognitive Concepts?

Given the amount of data generated by comparative psychologists suggesting differences in learning ability between different species, the trend toward interpreting invertebrate learning in terms of "representations," "cognitive maps," and other language borrowed from the vocabulary of human information processing seems to me unwarranted at this time. Consider, for instance, a recent report that interprets the simple choice behavior of a bumblebee as if these animals use "computational rules" and "cognitive architectures" (Real, 1991). As another example, Lloyd (1986) *prefers* to interpret the movement of a snail toward a light as the manipulation of the snails "inner representation." As a third example, consider the concept of honeybee "language." Although not generally acknowledged, von Frisch consistently placed quotation marks around the word *language*, thereby implying that bees may not truly possess a symbolic form of communication. Unfortunately, von Frisch's metaphor of bee "language" has been so carelessly paraphrased over the years that a generation of students and nonspecialists are convinced that bees possess a symbolic form of communication. In this regard, it is worth repeating von Frisch's (1962) own view of bee language:

> Before beginning the story I should like to emphasize the limitations of the language metaphor. The true comparative linguist is concerned with one of the subtlest products of man's powerfully developed thought processes. The brain of a bee is the size of a grass seed and is not made for thinking. The actions of bees are mainly governed by instinct. Therefore the student of even so complicated and purposeful an activity as the communication dance must remember that he is dealing with innate patterns, impressed on the nervous system of the insects over the immense reaches of time of their phylogenetic development. (p. 79)

As another example of the extension of cognitive concepts to invertebrate learning, let us consider how contemporary views of classical conditioning, developed with vertebrates, have affected our thinking about conditioning in invertebrates. For Pavlov and his contemporaries, classical conditioning involved the straightfoward association of two events occurring closely in time. As a result of such vertebrate experiments as blocking, overshadowing, and manipulating the contingency between the CS and US, higher vertebrates are now thought to detect and evaluate information embedded in the classical conditioning situation. (For a review of these and other studies supporting a cognitive interpretation of classical conditioning, see Mackintosh, 1983.) Some invertebrate learning studies have replicated these experiments with invertebrates (using CSs that are not neutral) and suggest that their preparations can tap into the analogous vertebrate processes (e.g., see Farley, 1987a, 1987b; Hawkins et al., 1986; Sahley et al., 1981).

It should be noted that there is no consensus in the vertebrate literature on whether or not simply pairing a CS and a US is a *sufficient* condition for producing classical conditioning. Wasserman (1989) and Papini and Bitterman (1990) both point out conceptual and experimental design flaws in the contingency view of classical conditioning. In addition, the trend to characterize vertebrate learning in terms of cognitive concepts is not without challenge (e.g., see Amsel, 1989). Even if it is concluded that some vertebrate species possess cognitive structures, it remains doubtful whether invertebrates also possess such structures. Blocking, overshadowing, and successive negative contrast in the honeybee, for example, can be understood without recourse to cognitive constructs (Couvillon & Bitterman, 1982, 1984). When attempting to interpret invertebrate behavior in terms of cognitive constructs, it is important to recall C. Lloyd Morgan's canon that we should look for the simplest explanation of a phenomenon. Consider, also, Tavolga's (1969) statement that "any attempt to interpret the behavior of an insect with concepts and methods based on human psychology would be as ridiculous as attempting a Freudian psychoanalysis of a cockroach" (p. 21).

Where Is the Study of Invertebrate Learning Today, and What Can You Contribute to It?

In closing, I do not wish to leave you with the impression that the study of invertebrate learning is mired in issues that can never be resolved—quite the contrary. There is a great need for competent researchers, and there are many opportunities to do first-rate research.

One of the more entertaining and instructive ways to identify trends in the invertebrate learning literature is to perform a computer literature search. In this section, the results of such a search performed on the *Medline* database are presented for the years 1966—the first year of the database—to 1991. The Medline database surveys 3,030 journals and is one of a number of readily accessible and widely used computer-based literature searches. We will see which invertebrate is "hot" and which is not, how the interest in learning and behavior declined or increased over the years, and in which animal(s) additional work should be directed. Three tables are provided, each listing 15 invertebrates, and indicating (a) the total number of experiments performed on that animal, (b) the total number of experiments that deal with learning, and (c) the total number of experiments that deal with behavior. The category of behavior includes appetitive behavior, consummatory behavior, and sexual behavior. The category of learning includes more than 40 terms such as *habituation*, *classical conditioning*, *operant conditioning*, and *avoidance*. The 15 invertebrates were selected primarily because of their importance in invertebrate research. The animals selected are ant, *Aplysia*, bee, *C. elegans*, crab, crayfish, *Drosophila*, grasshopper, *Helix*, *Hermissenda*, leech, *Limax*, lobster, *Pleurobranchaea*, and roach.

The results of this search highlight some of the problems that were discussed in chapter 2 regarding conducting such searches. First, databases are limited in the survey of the early literature. *Medline*, for instance, goes back only as far as 1966. Second, it should be recognized that *Medline*, like most databases, is continuously updated. Some journals, for instance, are added, others are dropped, key words are modified, and major descriptors are added, modified, or deleted. In addition, subject headings are often changed or added. These constant modifications should be kept in mind when reviewing the tables and figures that follow.

Findings

While searching the invertebrate literature, one is immediately struck by the lack of available information when compared to studies of vertebrate learning. Whether we compare the overall number of experiments or divide the overall number into behavior or learning experiments, the study of vertebrates far outstrips the interest in invertebrates. Figures 8-1 through 8-3 show the result of a computer literature search from 1966 to 1991, comparing (a) the total number of invertebrate and vertebrate studies, (b) the total number of invertebrate and vertebrate learning experiments, and (c) the total number of invertebrate and vertebrate behavior experiments, respectively. Several facts present themselves.

First, in terms of the overall number of experiments, studies of vertebrates far surpass the number of invertebrate experiments. Figure 8-1A shows that from 1966 to 1991, the number of invertebrate experiments rose from about 4,500 to 9,500. Figure 8-1B shows that the number of vertebrate experiments rose from almost 40,000 to about 90,000 papers during the same time period. In 1991 alone, there were over 74,000 more studies using vertebrates than invertebrates.

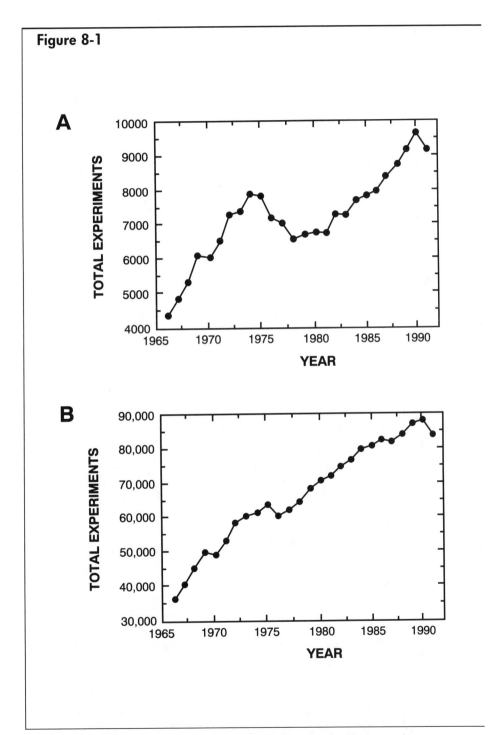

Figure 8-1

Total number of experiments for the years 1966 to 1991: (A) Invertebrate experiments; (B) Vertebrate experiments.

Figure 8-2

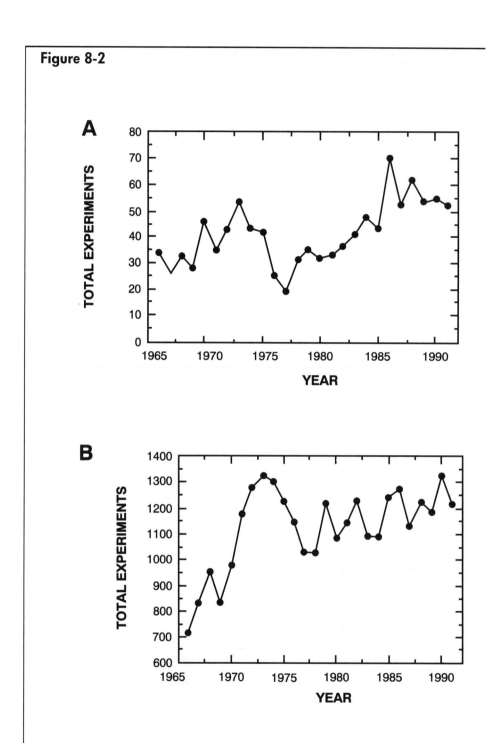

Total number of learning experiments for the years 1966 to 1991: (*A*) Invertebrate experiments; (*B*) Vertebrate experiments.

Figure 8-3

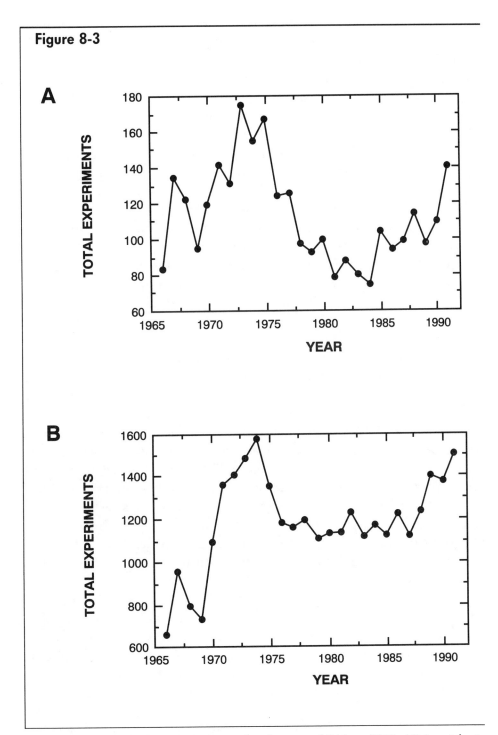

Total number of behavior experiments for the years 1966 to 1991: (A) Invertebrate experiments; (B) Vertebrate experiments.

Second, Figure 8-2A shows that the interest in invertebrate learning is surprisingly small, ranging from the 34 papers that appeared in 1966 to the 53 papers that appeared in 1991. There was a small increase in invertebrate learning studies from 1966 to 1975, followed by a decline lasting several years. Interest slowly returned from 1978 to 1987—possibly a result of the use of invertebrates in simple systems research—where it remains stable at about 50 papers per year. Despite the use of invertebrates in simple systems research, the study of invertebrate learning remains small—especially so when compared to the number of vertebrate learning experiments. Figure 8-2B shows that in 1966, there were 716 studies of vertebrate learning, which rose to 1,221 in 1991. The difference between invertebrate and vertebrate learning experiments in 1991, for example, was more than 1,100 papers.

Third, a comparison between Figure 8-2A and 8-3A suggests that there is more of an interest in studies of invertebrate behavior than in studies of learning. In 1991, for example, the database uncovered 53 studies of learning and 140 studies of behavior. This may reflect, however, nothing more than the fact that the database recognizes more "kinds" of behavior than "kinds" of learning. Alternatively, it may reflect the belief once held by many researchers that invertebrates could not learn, and even if they could, such learning was no more than a laboratory curiosity and not as important under natural conditions as other types of behavior such as that involved in social and reproductive activities. The large disparity between invertebrate behavior and learning studies also reflects the fact that it is easier to study the natural behavior of invertebrates than it is to devise clever tasks to reveal learning in the laboratory.

Figure 8-3A shows that the number of behavior experiments rose from 84 in 1966 to 140 by 1991. Even though the number of invertebrate behavior experiments is almost three times as many as that for learning experiments, it still pales in comparison with the number of vertebrate behavior experiments. Figure 8-3B shows that the number of behavior experiments using vertebrates rose from 656 in 1966 to almost 1,500 in 1991. In 1991 the difference between the number of invertebrate and vertebrate papers was more than 1,300. It is interesting to note that a comparison of the vertebrate data in Figure 8-2B and 8-3B reveals that throughout most of the years between 1966 and 1991, there was a close correspondence between the number of vertebrate studies of learning and behavior.

Figures 8-1 through 8-3 clearly illustrate that there is an appalling lack of experiments on invertebrate learning and behavior when compared to similar vertebrate experiments. Considering that invertebrates represent about 95 percent of all animal species, such a lack of attention is at best irresponsible, and considering the economic, medical, and environmental importance of invertebrate animals, dangerous.

Tables 8-2 through 8-4 extend the analysis to 15 individual invertebrates. These invertebrates were selected because of their popularity in research and contemporary interest.

Total experiments. As one might expect, the most popular invertebrate is the fruit fly, *Drosophila*. The number of experiments steadily rose from 224 in 1966 to more than 1,100 by 1991. In contrast, a steady decline can be seen in the study of the social insects (e.g., ants and bees) and somewhat in the use of crustaceans (e.g., crabs, crayfish, and lobsters). An increase in the number of experiments is readily apparent in molluscs (e.g., *Aplysia*, and *Hermissenda*) and other invertebrates actively involved in the search for the neural correlates of learning (e.g., *Helix*, leech, *Limax*, and *Pleurobranchaea*). Table 8-2 presents individual graphs showing the total number of experiments for each of the 15 invertebrates from 1966 to 1991.

Learning experiments. In terms of the number of learning experiments employing the 15 representative invertebrates, the results are quite disappointing. Since 1966, with the exception of *Aplysia*, bee, *Drosophila*, *Helix*, and *Hermissenda*, there are no invertebrates with 10 or more learning experiments in any given year. Interestingly, bees are no longer the dominant animal in invertebrate research. Interest in other invertebrates has dramatically increased relative to bee studies. From 1980 to 1991, 25 studies of learning in bees have been published. During the same time period, there were 140 studies of learning in *Aplysia*, 89 studies of learning in *Drosophila*, 72 studies of learning in *Helix*, and 56 studies of learning in *Hermissenda*.

Table 8-3 indicates the total number of learning experiments for each of the 15 invertebrates from 1966 to 1991. It is important to note that the "0" associated with a particular year does not necessarily indicate that no learning experiments were reported that year. It means that the database does not, at this time, detect any experiments. An example of this is several papers on learning in *Pleurobranchaea* that have appeared but were not detected by the database.

Behavior experiments. Interest in behavior experiments is found throughout most of the 15 invertebrates and is heaviest for *Aplysia*, *Drosophila*, and *Helix*. Table 8-4 indicates the total number of behavior experiments for each of the 15 invertebrates from 1966 to 1991. As was

Table 8-2

Total Invertebrate Experiments, 1966–1991

Year	Ant	Aplysia	Bee	C. elegans	Crab	Crayfish	Dro-sophila	Grass-hopper	Hermis-senda	Helix	Leech	Lobster	Limex	Pleuro-branchaea	Roach
1966	15	8	63	4	0	0	224	0	0	53	5	0	0	0	1
1967	23	15	74	1	1	1	290	0	1	44	9	0	1	0	0
1968	28	15	88	1	1	1	300	1	0	68	3	1	2	0	3
1969	15	18	95	2	0	3	310	1	1	59	12	1	5	0	1
1970	18	28	82	0	13	3	323	3	0	61	12	5	6	0	12
1971	34	28	126	1	63	50	355	34	0	79	2	37	4	0	53
1972	40	37	111	2	140	85	399	79	1	113	4	67	6	1	101
1973	31	30	101	1	155	113	442	87	2	98	6	61	5	2	105
1974	34	49	115	10	154	119	479	132	4	105	7	60	7	3	137
1975	46	55	116	10	160	101	521	102	5	164	4	84	3	1	116
1976	36	0	88	0	109	118	452	86	3	69	7	77	2	2	91
1977	28	33	106	0	95	89	397	100	0	73	1	56	9	2	84
1978	24	76	59	12	57	96	389	71	5	67	6	50	6	2	76
1979	13	102	62	15	80	80	465	87	4	90	4	61	4	0	64
1980	13	94	54	31	67	93	519	57	4	65	7	47	9	2	56
1981	13	100	43	32	50	87	526	84	3	85	5	61	4	0	65
1982	17	85	52	45	66	78	541	74	8	78	9	51	9	6	57
1983	8	92	66	46	51	87	519	92	6	75	8	58	10	9	55
1984	12	98	52	47	74	82	514	69	8	66	15	55	13	2	61
1985	16	99	43	58	71	73	682	97	7	69	10	36	9	6	73
1986	10	117	44	57	70	68	635	79	12	63	18	49	13	6	54
1987	6	115	45	71	58	86	694	68	10	79	26	56	13	2	63
1988	19	112	48	100	55	67	850	77	9	54	22	51	20	4	75
1989	15	131	60	107	65	79	1009	70	6	76	19	44	16	1	81
1990	19	105	55	127	41	76	1162	80	15	65	25	45	21	2	96
1991	15	104	54	106	58	78	1151	84	9	71	29	63	15	2	77

Table 8-3

Total Invertebrate Learning Experiments, 1966–1991

Year	Ant	Aplysia	Bee	C. elegans	Crab	Crayfish	Dro-sophila	Grass-hopper	Hermis-senda	Helix	Leech	Lobster	Limex	Pleuro-branchaea	Roach
1966	0	0	3	0	0	0	0	0	0	0	0	0	0	0	0
1967	1	1	5	0	0	0	4	0	0	0	0	0	0	0	0
1968	1	0	1	0	0	0	1	0	0	1	0	0	0	0	0
1969	1	0	0	0	0	0	1	0	0	0	0	0	0	0	0
1970	0	4	3	0	2	0	0	0	0	0	0	1	0	0	1
1971	2	2	1	0	0	4	1	1	0	1	0	3	0	0	2
1972	2	1	5	0	0	2	2	1	0	0	0	0	0	1	6
1973	2	2	10	0	4	4	2	0	1	0	0	1	0	0	6
1974	0	3	5	0	1	3	4	3	1	0	0	0	0	1	6
1975	0	2	5	0	2	1	3	3	0	0	0	0	1	1	2
1976	0	0	3	0	0	4	4	1	0	2	0	0	0	1	1
1977	0	0	1	0	1	1	2	2	1	0	0	0	0	0	0
1978	0	7	2	0	2	2	4	0	1	4	0	1	0	1	1
1979	1	12	4	0	0	1	2	1	1	5	0	0	1	0	0
1980	0	5	1	0	0	2	4	1	1	5	0	0	0	1	1
1981	1	9	1	0	0	1	4	1	1	6	0	0	1	0	2
1982	2	3	2	0	0	0	5	0	4	5	0	0	0	0	1
1983	0	12	3	0	0	1	6	4	3	5	0	0	1	1	0
1984	1	11	2	0	0	1	7	4	4	6	1	0	1	1	2
1985	0	7	1	0	0	1	9	0	4	7	0	0	0	5	1
1986	2	20	3	0	1	0	13	1	7	5	2	0	2	0	0
1987	0	16	3	0	0	0	7	0	7	10	0	1	0	0	1
1988	0	22	3	0	3	0	6	0	5	5	2	1	0	1	0
1989	1	17	1	0	1	0	8	0	5	7	0	0	2	0	0
1990	1	9	1	1	4	0	6	0	13	7	1	1	2	1	1
1991	1	9	4	1	3	1	14	0	2	4	1	3	0	0	0

Table 8-4

Total Invertebrate Behavior Experiments, 1966–1991

Year	Ant	Aplysia	Bee	C. elegans	Crab	Crayfish	Dro-sophila	Grass-hopper	Hermis-senda	Helix	Leech	Lobster	Limex	Pleuro-branchaea	Roach
1966	6	0	5	0	0	0	8	0	0	0	0	0	0	0	0
1967	4	0	10	0	0	0	18	0	0	0	0	0	0	0	0
1968	4	0	5	0	0	0	12	0	0	0	0	0	0	0	0
1969	3	0	5	0	0	0	3	0	0	0	0	0	0	0	0
1970	5	0	5	0	0	0	9	0	0	0	0	0	0	0	1
1971	6	2	10	0	2	0	13	3	0	0	0	2	0	0	2
1972	9	0	8	0	3	7	17	6	0	0	0	1	0	0	1
1973	5	0	8	0	4	0	24	4	1	0	0	0	0	0	7
1974	3	6	2	0	4	5	14	4	3	1	0	0	0	1	10
1975	7	1	9	0	3	3	23	2	0	0	0	1	0	0	9
1976	5	0	3	0	0	2	22	5	0	0	0	0	0	1	7
1977	4	1	3	0	2	3	20	4	0	1	0	0	0	1	3
1978	6	2	3	0	1	2	14	1	2	0	0	1	0	1	3
1979	1	9	4	1	1	1	20	3	1	2	0	0	0	1	1
1980	2	7	4	2	0	3	30	1	1	2	0	0	0	0	0
1981	0	7	2	2	0	3	15	1	0	2	0	0	0	1	3
1982	1	4	1	2	0	4	12	2	1	4	0	1	0	0	2
1983	0	8	2	1	0	1	11	1	1	2	0	0	0	1	0
1984	0	7	2	0	0	4	11	2	1	0	0	1	0	1	1
1985	2	3	1	0	1	1	18	4	1	1	1	2	0	0	5
1986	1	5	0	1	1	3	13	0	0	1	1	1	0	0	3
1987	0	4	2	0	1	2	25	1	1	2	2	0	0	0	2
1988	1	8	5	0	1	3	23	1	0	2	2	2	0	0	5
1989	1	11	1	0	2	3	15	0	0	2	0	0	0	0	2
1990	2	5	3	3	3	4	21	2	1	5	0	0	1	0	5
1991	3	12	6	2	5	2	19	1	0	11	3	2	0	0	4

the case for the learning experiments, note that the "0" associated with a particular year does not necessarily indicate that no learning experiments were reported, only that they were not detected.

Summary

One of the goals of this chapter was to encourage you to get away from the tendency to use such simplified terms as *classical* and *operant conditioning* and replace them with detailed classifications of behavioral methods. A second goal was to point out that there may be some very important differences between the learning of invertebrates and vertebrates. To encourage you to participate in invertebrate learning research, the results of a computer search were presented that identified invertebrates that are "hot" and those that can benefit from additional research.

Discussion Questions

- What are some of the general issues in the study of invertebrate learning?
- What kinds of inconsistencies exist in the definitions of learning phenomena as they apply to invertebrates?
- Look up one of the available behavioral taxonomies, and apply it to invertebrate research on classical, instrumental, and operant conditioning.
- Do you agree or disagree with the view that cognitive concepts can be applied to the learning of invertebrates?
- What are the similarities and differences between punishment and avoidance?
- Should the concept of operant behavior cover innate or reflexive behavior?
- Why is it important to include individual data in a research report?
- Using the results of the literature search presented in this chapter, identify some trends in invertebrate learning research.

Appendix A: Sources of Background Information

The articles cited in Table A-1 will enable you to obtain background material on the natural history and learning ability of many invertebrates.

Table A-1	
Review Materials	
Animal	**Source**
Ant	Vowles (1967); Alloway (1973); Hölldobler & Wilson (1990)
Aplysia	Kandel (1979)[a]; Farley & Alkon (1985)[a]; Carew & Sahley (1986)[a]; Byrne (1987)[a]; Schreurs (1989); Hawkins (1991)[a]
Bee	Wells (1973); Menzel, Erber, & Masuhr (1974); Gould (1986); Bitterman (1988); Goodman & Fisher (1991); Menzel, Hammer, Braun, Mauelshagen, & Sugawa (1991)[a]
Crab	Crothers (1967, 1968); Abramson & Feinman (1990b); Feinman, Abramson, & Forman (1990); Mantel (1992)
Crayfish	Krasne (1973); Goldman (1983)
Earthworm	Ratner (1965, 1967); Dyal (1973); Wyers et al. (1973)
Fruit fly	McGuire (1984); Tully (1984, 1991)[a]; Hirsch & Holliday (1988)[a]; Ricker, Hirsch, et al. (1986)
Hermissenda	Crow (1988)[a]; Carew & Sahley (1986)[a]; Farley & Alkon (1985)[a]; Byrne (1987)[a]; Schreurs (1989)
Housefly	McGuire (1984)
Limax	Carew & Sahley (1986)[a]; Farley & Alkon (1985)[a]; Davis (1986)
Lobster	Fine-Levy et al. (1988)

continues

Table A-1, continued

Animal	Source
Locust	Hoyle (1980)[a]; Carew & Sahley (1986)[a]; Farley & Alkon (1985)[a]; Byrne (1987)[a]
Nematode	Wharton (1986); Rankin, Beck, & Chiba (1990)
Planarian	Corning & Riccio (1970); Corning & Kelly (1973)
Pleurobranchaea	Carew & Sahley (1986)[a]; Farley & Alkon (1985)[a]; Byrne (1987)[a]; Davis (1986)[a]
Protozoan	Applewhite & Morowitz (1966)[a]; Corning & Von Burg (1973); Laybourn-Parry (1984); Wichterman (1986)
Roach	Alloway (1973); Eisenstein & Reep (1985)
General issues in invertebrate learning	Corning & Lahue (1972); Abraham, Palka, Peeke, & Willows (1972); Corning et al. (1976); Carew, Abrams, Hawkins, & Kandel (1984); Farley & Alkon (1985); Bullock (1986); Schreurs (1989)

Note. [a]Contains information about the underlying neuronal mechanisms.

Table A-2

Natural History and Culture Techniques

Animal	Source
Ant	Lutz, Welch, Goltsoff, & Needham (1937/1959); Hölldobler & Wilson (1990)
Aplysia	Kandel (1979); Berg (1983); Leonard & Lukowiak (1986)
Bee	Sammataro & Avitabile (1978)
Crab	Provenzano (1985)
Crayfish	Huxley (1880/1973)
Cricket	Gandwere (1960)
Earthworm	Lutz et al. (1937/1959); Edwards & Lofty (1977)
Fruit fly	Lutz et al. (1937/1959); Shorrocks (1972)
Housefly	West (1951)
Hermissenda	Berg (1983)
Hydra	Lenhoff (1983)
Jewel wasp	Barrass (1976)
Lobster	Berg (1983)
Locust	Lutz et al. (1937/1959)
Nematode	Wharton (1986)
Octopus	Berg (1983)
Planarian	McConnell (1967a)
Protozoa	Tartar (1961); Lee, Hunter, & Bovee, (1985); Ricci (1990)

Table A-3 provides a list of some of the apparatus discussed in chapter 3. Tips on constructing some of these apparatus and where to find parts and low-cost control equipment are available from my laboratory manual (Abramson, 1990).

Table A-3

Some Instrumentation Used in the Study of Invertebrate Learning

Type	Animal	Reference
\-	\-	Unautomated
Runway	Ant	Abramson et al. (1982)
	Earthworm	Reynierse & Ratner (1964); Kirk & Thompson (1967)[a]
	Roach	Longo (1970)
Maze	Ant	Schneirla (1933); Vowles (1964)
	Aplysia	Preston & Lee (1973)
	Bee	Menzel & Erber (1972)
	Crab	Datta et al. (1960)
	Crayfish	Capretta & Rea (1967)
	Beetle	Alloway (1972)
	Earthworm	Datta (1962); Keshavamurthy & Krishnamoorthy (1977)[a]; McManus & Wyers (1979)[a]; Rosenkoetter & Boice (1975)[a]
	Fruit fly	Dudai (1977); Drudge & Platt (1979)
	Lobster	Schöne (1961)
	Planarian	Best & Rubinstein (1962); Corning (1964); McConnell (1967)[a]
	Roach	Longo (1964)
Free-foraging situation	Bee	Couvillon & Bitterman (1980); Abramson (1986)
	Blow fly	Fukushi (1985)
	Grasshopper	Bernays & Wrubel (1985)
	Housefly	Fukushi (1983)
Proboscis situation	Bee	Bitterman et al. (1983); Smith et al. (1991)
	Blow fly	Akahane & Amakawa (1983)
	Housefly	Fukushi (1976)
\-	\-	Automated
Operant chamber	*Aplysia*	Downey & Jahan-Parwar (1972)
	Bee	Sigurdson (1981a, 1981b); Pessotti & Lignelli-Otero (1981)
	Crab	Abramson & Feinman (1990b)
	Crayfish	Olson & Strandberg (1979)
	Snail	Balaban & Chase (1989)

continues

Table A-3, continued

Type	Animal	Reference
		Automated
Shuttle box	Ant	Abramson, Collier, & Marcucella (1977); Abramson et al. (1982)
	Bee	Abramson (1986); Lee & Bitterman (1990a)
	Crab	Fernandez-Duque et al. (1992)
	Crayfish	Taylor (1971)
	Horseshoe crab	Makous (1969)
	Housefly	Leeming & Little (1977)
	Roach	Brown & Stroup (1988)
Running wheel	Earthworm	Marian & Abramson (1982)
	Fruit fly	DeJianne et al. (1985)
	Housefly	Miller et al. (1971)
	Roach	Ball (1972)
	Slug	Sokolove, Beiswanger, Prior, & Gelperin (1977)
Horridge task	Crab	Hoyle (1976); Dunn & Barnes (1981a, 1981b)
	Fruit fly	Booker & Quinn (1981); Mariath (1985)
	Locust	Forman (1984)
	Roach	Horridge (1962); Carrega & Huber (1985); Harris (1991); Eisenstein (1970)[a]
	Snail	Christoffersen et al. (1981)
Actograph	—	Brunner & Maldonado (1988)
	—	Kirkpatrick et al. (1991)
	—	Maldonado (1970)
Wind tunnel	—	Miller & Roelofs (1978)
	—	Sanders (1985)

Note. [a]Contains additional information about the use of the technique.

Appendix B: A Directory of Selected Invertebrate Researchers

Appendix B provides a partial directory of scientists engaged in invertebrate research. Before committing yourself to an invertebrate research program or technique, contact the scientist(s) in whose work you are interested. The scientists were selected because of their familiarity with a research program and/or extensive experience with a particular invertebrate.

The addresses of these and other scientists engaged in invertebrate research are readily available by consulting the membership directories of such scientific societies as the American Psychological Association, American Psychological Society, Federation of American Societies for Experimental Biology, International Brain Research Organization, Psychonomic Society, and the Society of Neuroscience. The addresses are, of course, also available from research articles. Unless the article is very recent, however, the address may no longer be current.

Thomas W. Abrams
Department of Biology
University of Pennsylvania
Leidy Laboratories
Philadelphia, PA 19104-6018
Major animal(s): Molluscs
Approach: Simple systems

Charles I. Abramson
Department of Psychology
Oklahoma State University
215 North Murray
Stillwater, OK 74078-0250
Major animal(s): Insects, crustaceans
Approach: Comparative

Daniel L. Alkon, MD
6701 Bonaventure Court
Bethesda, MD 20817
Major animal(s): Molluscs
Approach: Simple systems

Jelle Atema
Marine Biological Laboratory
Boston University Marine Program
Woods Hole, MA 02543
Major animal(s): Crustaceans
Approach: Simple systems

Harold L. Atwood
Department of Physiology
University of Toronto
Medical Sciences Building
Toronto, Ontario
Canada, M5S 1A8
Major animal(s): Crustaceans
Approach: Simple systems

M. E. Bitterman
Békésy Laboratory of
 Neurobiology
University of Hawaii
1993 East-West Road
Honolulu, HI 96822
Major animal(s): Insects
Approach: Comparative

John Byrne
Department of Neurobiology
 and Anatomy
University of Texas Medical
 School
P.O. Box 20708
Houston, TX 77225
Major animal(s): Molluscs
Approach: Simple systems

Thomas J. Carew, PhD
Department of Psychology
Yale University
New Haven, CT 06520
Major animal(s): Molluscs
Approach: Simple systems

Patricia A. Couvillon
Békésy Laboratory of
 Neurobiology
University of Hawaii
1993 East-West Road
Honolulu, HI 96822
Major animal(s): Insects
Approach: Comparative

Charles D. Derby
Department of Biology
Georgia State University
Atlanta, GA 30303
Major animal(s): Crustaceans
Approach: Simple systems

John F. Disterhoft
CMS Biology
Northwestern University Medical
 School
303 East Chicago Avenue
Chicago, IL 60611
Major animal(s): Insects
Approach: Simple systems

Yadin Dudai
Department of Neurobiology
Weizmann Institute of Science
76100, Rehovot, Israel
Major animal(s): Insects
Approach: Behavioral genetics

Joseph Farley
Program in Neuroscience
Indiana University
Psychology Building
Bloomington, IN 47405
Major animal(s): Molluscs
Approach: Simple systems

Jacqueline B. Fine-Levy
Department of Biology
Georgia State University
Atlanta, GA 30303
Major animal(s): Crustaceans
Approach: Simple systems

Louis E. Gardner
Department of Psychology
Creighton University
Omaha, NE 68178
Major animal(s): Annelids
Approach: Comparative

Alan Gelperin
Department of Molecular Biophysics
AT&T Bell Laboratories
Murray Hill, NJ 07974
Major animal(s): Molluscs
Approach: Simple systems

Robert D. Hawkins
Center for Neurobiology and
    Behavior
Columbia University
722 West 168th Street
New York, NY 10032
Major animal(s): Molluscs
Approach: Simple systems

Jerry Hirsch
Department of Psychology
University of Illinois
603 East Daniel Street
Champaign, IL 61820
Major animal(s): Insects
Approach: Behavioral genetics

John G. Hildebrand
ARL Division of Neurobiology
University of Arizona
611 Gould-Simpson Science Building
Tucson, AZ 85721
Major animal(s): Insects
Approach: Simple systems

Eric R. Kandel, MD
Center for Neurobiology and
    Behavior
Columbia University
722 West 168th Street
New York, NY 10032
Major animal(s): Molluscs
Approach: Simple systems

Franklin B. Krasne
Department of Psychology and Brain
    Research Institute
University of California, Los Angeles
Los Angeles, CA 90024
Major animal(s): Crustaceans
Approach: Simple systems

Ken Lukowiak
Neuroscience Research Group
Faculty of Medicine
University of Calgary
3330 Hospital Drive NW
Calgary, Alberta
Canada, T2N 4N1
Major animal(s): Molluscs
Approach: Ethological/simple systems

Hector Maldonado
Laboratorio de Fisiología del
    Comportamiento Animal
Dpto. Biología
Pabellón 2 - Ciudad Universitaria
(1428) Buenos Aires
Argentina
Major animal(s): Insects, crustaceans
Approach: Comparative/simple systems

Catharine H. Rankin
Department of Psychology
University of British Columbia
2136 West Mall
Vancouver, BC
Canada V6T 1Z4
Major animal(s): Nematodes
Approach: Behavioral genetics

Christine L. Sahley
Department of Biological Sciences
Purdue University
West Lafayette, IN 47907
Major animal(s): Molluscs
Approach: Simple systems

Brian H. Smith
Department of Entomology
Ohio State University
1735 Neil Avenue
Columbus, OH 43210-1220
Major animal(s): Insects
Approach: Ethological/simple systems

Ethel Tobach
The American Museum of Natural
   History
79th Street at Central Park West
New York, NY 10024
Major animal(s): Insects, molluscs
Approach: Comparative

Tim Tully
Beckman Neuroscience Center
Cold Spring Harbor Laboratory
Cold Spring Harbor, NY 11724
Major animal(s): Insects
Approach: Behavioral genetics

Everett J. Wyers
47 Shore Rd.
East Setauket, NY 11733
Major animal(s): Annelids
Approach: Comparative

A. O. D. Willows
University of Washington
Friday Harbor Laboratories
620 University Road
Friday Harbor, WA 98250
Major animals: Molluscs
Approach: Ethological/simple systems

Glossary

Before you can conduct an experiment on invertebrate learning, it is necessary to familiarize yourself with some of the basic terminology. Like any scientific discipline, the study of learning contains its own specialized terminology. Mastering such terminology is essential if you are to communicate effectively with colleagues and correctly interpret the results from research reports. Those readers who are psychologists will find much of this vocabulary familiar. This is not by chance. The study of learning is shared by many disciplines, including biochemistry, neuroscience, and entomology. Nevertheless, much of the vocabulary of invertebrate learning is taken from the vocabulary of the psychologist. This is due primarily to the influence of such pioneers as Herrmann Ebbinghaus, Edward L. Thorndike, Ivan P. Pavlov, John B. Watson, Theodore C. Schneirla, Clark L. Hull, Edward C. Tolman, and B. F. Skinner. Moreover, many of the those who followed and made significant contributions to the study of invertebrate learning, such as M. E. Bitterman, Jerry Hirsch, Stanley Ratner, James McConnell, E. R. John, Philip Applewhite, W. C. Corning, and Frank Krasne were trained as psychologists.

Activity wheel (also known as a *running wheel* or *revolving drum*): A cage built in the shape of a wheel, used to measure the movement of an organism.

Actograph: A family of devices used to measure the activity of an organism. Actographs come in many shapes and sizes. Movements are recorded using a variety of methods such as photoelectric, ultrasonic, and mechanical.

Alpha conditioning: A form of associative learning in which a nonneutral stimulus is paired with a response-producing stimulus (US). The difference between alpha conditioning and classical conditioning is that the conditioned responses arise from an innate connection.

Amplitude of response: In nonassociative and associative learning, the difference between the preconditioning or baseline level of a response and the value of the response measured for all trials during training. Amplitude of response is one of the most common response measures found in studies of sensitization, habituation, and classical conditioning.

Analogy: Behavioral and/or anatomical similarities between two or more functionally similar species, resulting from factors other than common ancestry. For example, species not closely related may develop similar traits because their environments are similar. Such comparisons between species tend to be general, superficial, and functional.

Associative learning: A form of behavioral modification involving the association of two or more events such as between two stimuli, or between a stimulus and a response. Associative learning can be modeled using classical conditioning, instrumental conditioning, and operant conditioning techniques.

Avoidance conditioning: A conditioning situation in which a response postpones or omits the presentation of an aversive event.

Backward conditioning: In classical conditioning, a situation in which the onset of the unconditioned stimulus precedes, rather than follows, the onset of the conditioned stimulus.

Behavioral genetics: The field of research that attempts to identify the effects of genes on the expression of behavior and psychological traits.

Classical conditioning (also known as *Pavlovian conditioning*): A form of associative learning introduced by Pavlov in which a previously neutral (conditioned) stimulus is paired with a response-producing (unconditioned) stimulus. Through the pairing of the conditioned stimulus with the unconditioned stimulus, the conditioned stimulus becomes endowed with the power to evoke a response (the conditioned response) on virtually every trial.

Comparative method: The common name for a research strategy that attempts to seek similarities and differences to develop a more complete picture of an anatomical feature or behavior.

Compound conditioning: In classical conditioning, a situation in which the conditioned stimulus is composed of two or more discrete components. Thus, a compound conditioned stimulus for a honeybee might be composed of a petri dish containing a splash of color and a scent.

Conditioned response (CR): In classical conditioning, a response elicited by the conditioned stimulus as a result of pairing the conditioned stimulus with an unconditioned stimulus. The CR arises from associative learning rather than a heightened sensitivity to stimuli in general.

Conditioned stimulus (CS): In classical conditioning, the originally neutral stimulus that comes to elicit a conditioned response.

Control group: In an experiment, those subjects whose responses are used as a standard for comparison; their responses are compared with those of the experimental group(s).

Countercurrent situation: An olfactory conditioning device for training large numbers of fruit flies simultaneously.

Delayed conditioning: In classical conditioning, a situation in which the conditioned stimulus begins before the unconditioned stimulus (which is delayed) and remains until the onset of the unconditioned stimulus.

Dependent variable: The measured response in an experiment.

Discrimination: In classical, instrumental, and operant conditioning, the tendency established through training to respond to one stimulus but not another.

Escape conditioning: A type of instrumental and operant conditioning situation in which an aversive stimulus is turned off by an appropriate response.

Ethogram: A behavioral profile often obtained under natural conditions.

Ethology: The field of research that studies the behavior patterns of animals under natural conditions.

Experimental group: In an experiment, a group of subjects whose behavior is observed during the manipulation of an independent variable. Their responses are compared to those of a control group.

Extinction: In classical conditioning, the removal of the correlation between conditioned stimulus and unconditioned stimulus. In instrumental and operant conditioning, the removal of the correlation between response and reward. In most cases, this is accomplished by not presenting the unconditioned stimulus or reward.

Food preference procedure: A classical conditioning technique in which the conditioned stimulus is taste and the unconditioned stimulus is poison or food.

Free-flying procedure (also known as the *von Frisch technique*): An instrumental technique in which honeybees shuttle back and forth from a living area to the experimental area, where they take sucrose from targets distinguished by color, odor, and/or position.

Generalization: In classical, instrumental, and operant conditioning, the tendency to respond to stimuli that have not been paired with an unconditioned stimulus or reward. The strength of the response increases with the degree of similarity.

Habituation: A nonassociative type of learning in which there is a decrease in the strength of a response to a stimulus produced by repeated presentations of that stimulus.

Homology: Behavioral and/or anatomical similarities between two or more species, driving from a common evolutionary ancestor. Such comparisons between species tend to be close, detailed, and structural.

Horridge procedure (also known as *leg position learning* or *position learning*): An instrumental conditioning technique in which movement of an appendage beyond a specified position is rewarded.

Independent variable: A condition manipulated by the experimenter in order to determine the effect of such manipulation on the dependent variable.

Instrumental response: In instrumental conditioning, a response that is instrumental in obtaining a reward.

Latency of response: In nonassociative and associative learning, the time between the onset of a conditioned stimulus or discriminative stimulus and the beginning of the response. Response latency is one of the most common response measures found in studies of instrumental learning.

Learning: A relatively permanent change in an organism's behavior as a result of experience.

Lever press apparatus (also known as a *Skinner box*, *operant chamber*, or *automated problem box*): An environment in which an animal learns to manipulate some device to obtain a reward.

Magnitude of response: In nonassociative and associative learning, the difference between the preconditioning or baseline level of a response and the value of the response measured only on those trials in which a difference exists. Response magnitude is one of the most common response measures found in studies of sensitization, habituation, and classical conditioning.

Maze: A seminatural environment consisting of a number of pathways and choice points.

Naturalistic observation: Careful observation of events that are not manipulated.

Nonassociative learning: A form of behavioral modification involving the association of one event, such as when the repeated presentation of a stimulus leads to an alteration of the strength or probability of a response. Nonassociative learning can be modeled using habituation and sensitization techniques.

Operant: In operant conditioning, all responses having a common effect on the environment. An operant is independent of the topography of the response. The critical factor is that the result of the action influences the environment in the same way. Thus, a crab can press a lever for food with its right claw, its left claw, or with both claws, firmly or softly.

Performance: The behavior exhibited by an organism. It may or may not reflect what has been learned.

Phylogenetic scale: Supposed representation of the evolutionary relationships of a collection of organisms.

Proboscis conditioning (also known as *olfactory conditioning* or *proboscis extension reflex conditioning*): A classical conditioning technique in which an olfactory stimulus serves as the conditioned stimulus that is paired with a sucrose feeding.

Pseudoconditioning: In classical conditioning, instrumental conditioning, and operant conditioning, any "conditioning" that is shown not to have been dependent upon the correlation between the conditioned stimulus and unconditioned stimulus or upon the correlation between the response and reward.

Punishment conditioning: A type of instrumental and operant situation in which the presentation of a stimulus following a response results in a decrease in the future probability of the response.

Rate of response: In operant conditioning, the average number of operant responses emitted over a period of time. For example, if a bee pressed a lever 50 times in a 10-minute experiment, the response rate would be 5 per minute. Response rate is the most widely used measure of behavior in operant conditioning experiments.

Reflex: An unlearned response to a specific stimulus.

Runway: A maze without blind alleys.

Sensitization: A nonassociative type of learning in which there is an increase in the strength of a response to a stimulus, produced by repeated presentations of that stimulus.

Shuttle box (also known as a *choice chamber* or *double chamber*): A chamber consisting of two or more compartments that requires an organism to move from one compartment to another.

Simple systems strategy: The field of research that exploits animals with relatively well-understood nervous systems in the quest to understand the neural and biochemical basis of behavior.

Simultaneous conditioning: A type of classical conditioning in which the conditioned stimulus and unconditioned stimulus are presented at precisely the same time.

Spontaneous recovery: In habituation, sensitization, classical, instrumental, and operant conditioning, the reappearance of an extinguished response following a rest.

Taste conditioning: A classical conditioning situation in which taste serves as the conditioned stimulus.

Temporal conditioning: In classical conditioning, a situation in which the unconditioned stimulus is presented at prearranged intervals without a conditioned stimulus.

Trace conditioning: In classical conditioning, a situation in which a conditioned stimulus is presented and terminated prior to the introduction of the unconditioned stimulus.

Unconditioned response (UR): In classical conditioning, the response elicited by an unconditioned stimulus without prior training.

Unconditioned stimulus (US): In classical conditioning, the stimulus that elicits an unconditioned response.

Withdrawal conditioning: A family of techniques in which the conditioned response is a contraction of the whole animal or of an isolated portion of the animal.

References

Abby-Kalio, N. J. (1989). Learning and the role of cue reinforcement in the shell cracking behaviour of the shore crab: *Carcinus maenas* (L). *Behavioural Processes*, *18*, 141–153.

Abraham, F., Palka, J., Peeke, H., & Willows, A. O. D. (1972). Model neural systems and strategies for the neurobiology of learning. *Behavioral Biology*, *7*, 1–24.

Abramson, C. I. (1981). Passive avoidance in the California harvester ant. *Journal of General Psychology*, *104*, 29–40.

Abramson, C. I. (1986). Aversive conditioning in honeybees (*Apis mellifera*). *Journal of Comparative Psychology*, *100*, 108–116.

Abramson, C. I. (1990). *Invertebrate learning: A laboratory manual and source book*. Washington, DC: American Psychological Association.

Abramson, C. I., Armstrong, P. M., Feinman, R. A., & Feinman, R. D. (1988). Signaled avoidance learning in the eye withdrawal reflex of the green crab. *Journal of the Experimental Analysis of Behavior*, *50*, 483–492.

Abramson, C. I., & Bitterman, M. E. (1986a). Latent inhibition in honeybees. *Animal Learning and Behavior*, *13*, 184–189.

Abramson, C. I., & Bitterman, M. E. (1986b). The US–preexposure effect in honeybees. *Animal Learning and Behavior*, *14*, 374–379.

Abramson, C. I., Collier, D. M., Marcucella, H. (1977). An aversive conditioning unit for ants. *Behavior Research Methods & Instrumentation*, *9*, 505–507.

Abramson, C. I., & Feinman, R. D. (1987). Operant punishment of eye elevation in the green crab, *Carcinus maenas*. *Behavioral and Neural Biology*, *48*, 259–277.

Abramson, C. I., & Feinman, R. D. (1988). Classical conditioning of the eye withdrawal reflex in the green crab. *Journal of Neuroscience*, *8*, 2907–2912.

Abramson, C. I., & Feinman, R. D. (1990a). Lever-press conditioning in the crab. *Physiology & Behavior*, *48*, 267–272.

Abramson, C. I., & Feinman, R. D. (1990b). Operant conditioning in the crab. In K. Wiese & J. Tautz (Eds.), *Crustacean pioneer systems in neurobiology* (pp. 207–214). Berlin: Birkhauser-Verlag.

Abramson, C. I., Miler, J., & Mann, D. W. (1982). An olfactory shuttle box and runway for insects. *Journal of Mind and Behavior*, *3*, 151–159.

Akahane, R., & Amakawa, T. (1983). Stable and unstable phase of memory in classically conditioned fly, *Phormia regina*: Effects of nitrogen gas anaesthesia and cycloheximide injection. *Journal of Insect Physiology*, *29*, 331–337.

Alexander, Jr., J., Audesirk, T. E., & Audesirk, G. J. (1984). One-trial reward learning in the snail *Lymnea stagnalis*. *Journal of Neurobiology*, *15*, 67–72.

Alkon, D. L. (1974). Associative training of *Hermissenda*. *Journal of General Physiology, 64*, 70–84.

Alkon, D. L. (1975). Neural correlates of associative training in *Hermissenda*. *Journal of General Physiology, 65*, 46–56.

Alkon, D. L. (1983). Learning in a marine snail. *Scientific American, 249*, 70–84.

Alkon, D. L. (1987). *Memory traces in the brain*. Cambridge, England: Cambridge University Press.

Alkon, D. L. (1989). Memory storage and neural systems. *Scientific American, 26*, 42–50.

Alloway, T. M. (1969). Effects of low temperature upon acquisition and retention in the grain beetle (*Tenebrio molitor*). *Journal of Comparative and Physiological Psychology, 69*, 1–8.

Alloway, T. M. (1972). Retention of learning through metamorphosis in the grain beetle, *Tenebrio molitor*. *American Zoologist, 12*, 471–477.

Alloway, T. M. (1973). Learning in insects except *Apoidea*. In W. C. Corning, J. A. Dyal, & A. O. D. Willows (Eds.), *Invertebrate learning: Vol. 2. Arthropods and gastropod mollusks* (pp. 131–171). New York: Plenum.

Amsel, A. (1972). Inhibition and mediation in classical, Pavlovian and instrumental conditioning. In R. Boakes & S. Halliday (Eds.), *Inhibition and learning* (pp. 275–299). New York: Academic Press.

Amsel, A. (1989). *Behaviorism, neobehaviorism and cognitivism in learning theory: Historical and contemporary perspectives*. Hillsdale, NJ: Lawrence Erlbaum Associates.

Appleton, T., & Wilkens, J. L. (1990). Habituation, sensitization and the effect of serotonin on the eyestalk withdrawal reflex of *Cancer magister*. *Comparative Biochemistry and Physiology, 97A*, 159–163.

Applewhite, P. B. (1972). The flow of ions in learning and memory. *Journal of Theoretical Biology, 36*, 419–423.

Applewhite, P. B., & Gardner, F. (1971). A theory of protozoan habituation learning. *Nature, 230*, 285–287.

Applewhite, P. B., & Morowitz, H. J. (1966). The micrometazoa as model systems for studying the physiology of memory. *Yale Journal of Biology and Medicine, 39*, 90–105.

Applewhite, P. B., & Morowitz, H. J. (1967). Memory and the microinvertebrates. In W. C. Corning & S. C. Ratner (Eds.), *The chemistry of learning: Invertebrate research* (pp. 329–340). New York: Plenum.

Aranda, L. C., Fernandez, O. P., Celume, E. S., & Luco, J. V. (1968). The influence of the four first anterior ganglia on learned behavior in *Lumbricus terrestris*. *Physiology & Behavior, 3*, 753–756.

Aranda, L. C., & Luco, J. V. (1969). Further studies of an electric correlate to learning. Experiments in an isolated ganglion. *Physiology & Behavior, 4*, 133–137.

Aronson, L. R., Tobach, E., Rosenblatt, J. S., & Lehrman, D. S. (Eds.). (1972). *Selected writings of T. C. Schneirla*. San Francisco: W. H. Freeman.

Atkinson, R. J. A., Bailey, H., & Naylor, E. (1974). Some laboratory methods for recording and displaying temporal patterns of locomotor activity in marine animals. *Marine Behaviour and Physiology, 3*, 59–70.

Audesirk, T., Alexander, J., Jr., Audesirk, G., & Moyer, C. (1982). Rapid, nonaversive conditioning in a gastropod. I. Effects of age and motivation. *Behavioral and Neural Biology, 36*, 379–390.

Baerends, G. P. (1988). Ethology. In R. C. Atkinson, R. J. Herrnstein, G. Lindzey, & R. D. Luce (Eds.), *Steven's handbook of experimental psychology* (2nd ed.; pp. 765–830). New York: Wiley.

Baker, J. R. (1952). *Abraham Trembley of Geneva: Scientist and philosopher 1710–1784.* London: Edward Arnold and Co.

Balaban, P. M., & Chase, R. (1989). Self-stimulation in snails. *Neuroscience Research Communications, 4,* 139–146.

Balderrama, N., & Maldonado, H. (1971). Habituation of the deimatic response in the mantid (*Stagmatoptera biocellata*). *Journal of Comparative and Physiological Psychology, 75,* 98–106.

Ball, H. J. (1972). A system for recording activity of small insects. *Journal of Economic Entomology, 65,* 129–132.

Barrass, R. (1976). Rearing jewel wasps *Mormoniella vitripennis* (Walker) and their use in teaching biology. *Journal of Biology Education, 10,* 119–126.

Batson, J. D., Hoban, J. S., & Bitterman, M. E. (1992). Simultaneous conditioning in honeybees (*Apis mellifera*). *Journal of Comparative Psychology, 106,* 114–119.

Baxter, C. (1957). Habituation of the roach to puffs of air. *Anatomical Record, 128,* 52.

Beach, F. A. (1950). The snark was a boojum. *American Psychologist, 5,* 115–124.

Becker, Jr., A. J., & Valinski, W. A. (1981). A method for behavior modification in aquatic crustaceans. *Proceedings of the West Virginia Academy of Science, 53,* 50–54.

Bell, W. J., & Tobin, T. R. (1982). Chemo-orientation. *Biological Review, 57,* 219–260.

Benzer, S. (1967). Behavioral mutants of *Drosophila* isolated by countercurrent distribution. *Proceedings of the National Academy of Science, 58,* 1112–1119.

Benzer, S. (1973). Genetic dissection of behavior. *Scientific American, 229,* 24–37.

Berg, C. J. (Ed.). (1983). *Culture of marine invertebrates: Selected readings.* Stroudsburg, PA: Hutchinson-Ross Publishing Co.

Bernays, E. A., & Lee, J. C. (1988). Food aversion learning in the polyphagous grasshopper *Schistocerca americana. Physiological Entomology, 13,* 131–138.

Bernays, E. A., & Wrubel, R. P. (1985). Learning by grasshoppers: Association of colour/light intensity with food. *Physiological Entomology, 10,* 359–369.

Best, J. B., & Rubinstein, I. (1962). Maze learning and associated behavior in planaria. *Journal of Comparative and Physiological Psychology, 55,* 560–566.

Bindra, D. (1976). *A theory of intelligent behavior.* New York: Wiley.

Bitterman, M. E. (1960). Toward a comparative psychology of learning. *American Psychologist, 15,* 704–712.

Bitterman, M. E. (1962). Techniques for the study of learning in animals: Analysis and classification. *Psychological Bulletin, 59,* 81–93.

Bitterman, M. E. (1965). The CS-US interval in classical and avoidance conditioning. In W. F. Prokasy (Ed.), *Classical conditioning* (pp. 1–19). New York: Appleton-Century-Crofts.

Bitterman, M. E. (1965). Phyletic differences in learning. *American Psychologist, 20,* 396–410.

Bitterman, M. E. (1967a). The evolution of intelligence. In J. L. McGaugh, N. M. Weinberger, & R. E. Whalen (Eds.), *Readings from Scientific American: Psychobiology* (pp. 150–157). San Francisco: W. H. Freeman.

Bitterman, M. E. (1967b). Learning in animals. In H. Helson & W. Bevan (Eds.), *Contemporary approaches to psychology* (pp. 139–179). Princeton, NJ: D. Van Nostrand and Co.

Bitterman, M. E. (1975). The comparative analysis of learning. *Science, 188,* 699–709.

Bitterman, M. E. (1976). Incentive contrast in honey bees. *Science, 192,* 380–382.

Bitterman, M. E. (1979). Historical introduction. In M. E. Bitterman, V. M. Lolordo, J. B. Overmier, & M. E. Rashotte (Eds.), *Animal learning: Survey and analysis* (pp. 1–24). New York: Plenum.

Bitterman, M. E. (1988). Vertebrate-invertebrate comparisons. In H. J. Jerison & I. Jerison (Eds.), *Intelligence and evolutionary biology* (pp. 251–276). Berlin: Springer-Verlag.

Bitterman, M. E., Menzel, R., Fietz, A., & Schäfer, S. (1983). Classical conditioning of proboscis extension in honeybees (*Apis mellifera*). *Journal of Comparative Psychology, 97,* 107–119.

Block, R. A., & McConnell, J. V. (1967). Classically conditioned discrimination in the planarian, *Dugesia dorotocephala. Nature, 215,* 1465–1466.

Blue, J. (1976). Effect of anterior ganglia removal on phototaxis in the earthworm (*Lumbricus terrestris*). *Bulletin of the Psychonomic Society, 7,* 257–259.

Boe, E. R., & Church, R. M. (Eds.). (1968). *Punishment: Issues and experiments.* New York: Appleton-Century-Crofts.

Booker, R., & Quinn, W. G. (1981). Conditioning of leg position in normal and mutant *Drosophila. Proceeding of the National Academy of Science, 78,* 3940–3944.

Boulis, N., & Sahley, C. L. (1988). A behavioral analysis of habituation and sensitization of shorting in the semi-intact leech. *Journal of Neuroscience, 8,* 4621–4627.

Braun, G., & Bicker, G. (1992). Habituation of an appetitive reflex in the honeybee. *Journal of Neurophysiology, 67,* 588–598.

Brown, H. M., Dustman, R. E., & Beck, E. C. (1966). Sensitization in planaria. *Physiology and Behavior, 1,* 305–308.

Brown, B. M., & Noble, E. P. (1967). Cycloheamide and learning in the isolated cockroach ganglion. *Brain Research, 6,* 363–366.

Brown, G. E., Hughes, G. D., & Jones, A. A. (1988). Effects of shock controllability on subsequent aggressive and defensive behaviors in the cockroach (*Periplaneta americana*). *Psychological Reports, 63,* 563–569.

Brown, G. E., & Stroup, K. (1988). Learned helplessness in the cockroach (*Periplaneta americana*). *Behavioral and Neural Biology, 50,* 246–250.

Bruner, J., & Tauc, L. (1965). Long-lasting phenomena in the molluscan nervous system. *Symposium Society of Experimental Biologists, 20,* 457–475.

Brunner, D., & Maldonado, H. (1988). Habituation in the crab *Chasmagnathus granulatus*: Effect of morphine and naloxone. *Journal of Comparative Physiology A, 162,* 687–694.

Buchanan, G. M., & Bitterman, M. E. (1988). Learning in honeybees as a function of amount and frequency of reward. *Animal Learning & Behavior, 16,* 247–255.

Buchanan, G. M., & Bitterman, M. E. (1989). Learning in honeybees as a function of amount of reward: Tests of the equal-asymptote assumption. *Animal Learning & Behavior, 17,* 475–480.

Buchsbaum, R. (1938). *Animals without backbones.* Chicago: University of Chicago Press.

Buchsbaum, R., Buchsbaum, M., Pearse, J., & Pearse, V. (1987). *Animals without backbones* (3rd ed.). Chicago: University of Chicago Press. (Original work published 1938)

Buerger, A. A., Eisenstein, E. M., & Reep, R. L. (1981). The yoked control in instrumental avoidance conditioning: An empirical and methodological analysis. *Physiological Psychology, 9,* 351–353.

Bullock, T. H. (1986). "Simple" model systems need comparative studies: Differences are as important as commonalities. *Trends in Neuroscience, 9,* 470–472.

Bullock, T. H., & Horridge, G. A. (1965). *Structure and function in the nervous systems of invertebrates: Vols. 1 & 2.* San Francisco: W. H. Freeman.

Bullock, T. H., Orkand, R., & Grinnell, A. (1977). *Introduction to nervous systems.* San Francisco: W. H. Freeman.

Bullock, T. H., & Quarton, G. C. (1966). Simple systems for the study of learning mechanisms. *Neurosciences Research Program Bulletin, 4*, 105–233.

Byrne, J. H. (1987). Cellular analysis of associative learning. *Physiological Review, 67*, 329–439.

Byrne, J. H. (1992). *Aplysia*: Classical conditioning and operant conditioning. In L. R. Squire (Ed.), *Encyclopedia of learning and memory* (pp. 44–47). New York: Macmillan.

Cadwallader, T. C. (1984). Neglected aspects of the evolution of American comparative and animal psychology. In G. Greenberg & E. Tobach (Eds.), *Behavioral evolution and integrative levels* (pp. 15–48). Hillsdale, NJ: Lawrence Erlbaum Associates.

Camhi, J. M. (1984). *Neuroethology: Nerve cells and the natural behavior of animals*. Sunderland, MA: Sinauer.

Campbell, C. B. G., & Hodos, W. (1991). The *Scala Naturae* revisited: Evolutionary scales and anagenesis in comparative psychology. *Journal of Comparative Psychology, 105*, 211–221.

Capretta, P. C., & Rea, R. (1967). Discrimination reversal learning in the crayfish. *Animal Behaviour, 15*, 6–7.

Carew, T. J. (1992). *Aplysia*: Development of processes underlying learning. In L. R. Squire (Ed.), *Encyclopedia of learning and memory* (pp. 51–56). New York: Macmillan.

Carew, T. J., Abrams, T. W., Hawkins, R. D., & Kandel, E. R. (1984). The use of simple invertebrate systems to explore psychological issues related to associative learning. In D. L. Alkon & J. Farley (Eds.), *Primary neural substrates of learning and behavioral change* (pp. 169–183). Cambridge: Cambridge University Press.

Carew, T. J., & Kupfermann, I. (1974). The influence of different natural environments on habituation in *Aplysia californica*. *Behavioral Biology, 12*, 339–345.

Carew, T. J., Pinsker, H. M., & Kandel, E. R. (1972). Longterm habituation of a defensive withdrawal reflex in *Aplysia*. *Science, 175*, 451–454.

Carew, T. J., & Sahley, C. L. (1986). Invertebrate learning and memory: From behavior to molecules. *Annual Review of Neuroscience, 9*, 435–487.

Carew, T. J., Walters, E. T., & Kandel, E. R. (1981). Classical conditioning in a simple withdrawal reflex in *Aplysia californica*. *Journal of Neuroscience, 1*, 1426–1437.

Carrega, L. M., & Huber, I. (1985). Learning and memory in the American cockroach: An apparatus to identify learning- and memory-deficient mutants. *Bulletin of the New Jersey Academy of Science, 30*, 49.

Carson, R. (1962). *Silent spring*. Boston: Houghton Mifflin.

Castellucci, V. (1992). *Aplysia*: Molecular basis of long-term sensitization. In L. R. Squire (Ed.), *Encyclopedia of learning and memory* (pp. 47–51). New York: Macmillan.

Catania, A. C. (1973). The concept of the operant in the analysis of behavior. *Behaviorism, 1*, 103–116.

Chen, W. Y., Aranda, L. C., & Luco, J. V. (1970). Learning and long- and short-term memory in cockroaches. *Animal Behaviour, 18*, 725–732.

Christoffersen, G. R. J., Frederiksen, K., Johansen, J., Kristensen, B. I., & Simonsen, L. (1981). Behavioural modification of the optic tentacle of *Helix pomatia*: Effect of puromycin, Activity of S-100. *Comparative Biochemistry and Physiology, 68A*, 611–624.

Church, R. M. (1964). Systematic effect of random error in the yoked control design. *Psychological Bulletin, 62*, 122–131.

Church, R. M., & Lerner, N. D. (1976). Does the headless roach learn to avoid? *Physiological Psychology, 4*, 439–442.

Clark, R. B. (1960a). Habituation of the polychaete *Nereis* to sudden stimuli. I. General properties of the habituation process. *Animal Behaviour, 8*, 82–91.

Clark, R. B. (1960b). Habituation of the polychaete *Nereis* to sudden stimuli. II. Biological significance of habituation. *Animal Behaviour, 8,* 92–103.

Cole, K. S. (1949). Dynamic electrical characteristics of the squid axon membrane. *Archives des Sciences Physiologiques, 3,* 253–256.

Cole, K. S. (1968). *Membranes, ions and impulses.* Berkeley: University of California Press.

Cook, A. (1971). Habituation in a freshwater snail (*Limnaea stagnalis*). *Animal Behaviour, 19,* 463–474.

Cook, D. G., & Carew, T. J. (1986). Operant conditioning of head-waving in *Aplysia. Proceedings of the National Academy of Sciences, 83,* 1120–1124.

Cook, D. G., & Carew, T. J. (1989a). Operant conditioning of head-waving in *Aplysia.* I. Identified muscles involved in the operant response. *Journal of Neuroscience, 9,* 3097–3106.

Cook, D. G., & Carew, T. J. (1989b). Operant conditioning of head-waving in *Aplysia.* II. Contingent modification of electromyographic activity in identified muscles. *Journal of Neuroscience, 9,* 3107–3114.

Cook, D. G., & Carew, T. J. (1989c). Operant conditioning of head-waving in *Aplysia.* III. Cellular analysis of possible reinforcement pathways. *Journal of Neuroscience, 9,* 3115–3122.

Copeland, M. (1930). An apparent conditioned response in *Nereis virens. Journal of Comparative Psychology, 10,* 339–354.

Corning, W. C. (1964). Evidence of right-left discrimination in planarians. *Journal of Psychology, 58,* 131–139.

Corning, W. C. (1971). Conditioning and "transfer" of training in a colonial ciliate: A summary of the work of N. N. Plavilstchikov. *Journal of Biological Psychology, 13,* 39–41.

Corning, W. C., Dyal, J. A., & Lahue, R. (1976). Intelligence: An invertebrate perspective. In R. B. Masterton, W. Hodos, & H. Jerison (Eds.), *Evolution, brain, and behavior: Persistent problems* (pp. 215–263). Hillsdale, NJ: Lawrence Erlbaum Associates.

Corning, W. C., Dyal, J. A., & Willows, A. O. D. (Eds.). (1973–1975). *Invertebrate learning: Vols. 1–3.* New York: Plenum.

Corning, W. C., & Kelly, S. (1973). Platyhelminthes: The turbellarians. In W. C. Corning, J. A. Dyal, & A. O. D. Willows (Eds.), *Invertebrate learning: Vol. 1. Protozoans through annelids* (pp. 171–224). New York: Plenum.

Corning, W. C., & Lahue, R. (1972). Invertebrate strategies in comparative learning studies. *American Zoologist, 12,* 455–469.

Corning, W. C., & Ratner, S. C. (Eds.). (1967). *Chemistry of learning.* New York: Plenum.

Corning, W. C., & Riccio, D. (1970). The planarian controversy. In W. Bryne (Ed.), *Molecular approaches to learning and memory* (pp. 107–150). New York: Academic Press.

Corning, W. C., & Von Burg, R. (1973). Protozoa. In W. C. Corning, J. A. Dyal, & A. O. D. Willows (Eds.), *Invertebrate learning: Vol. 1. Protozoans through annelids* (pp. 49–122). New York: Plenum.

Costanzo, D. J., & Cox, W. G. (1971). Habit reversal improvement in crayfish. *Journal of Biological Psychology, 13,* 11–12.

Costanzo, D. J., Rudolph, G. R., & Cox, W. (1972). Social status and habit reversal learning in crayfish. *Journal of Biological Psychology, 14,* 30–32.

Couvillon, P. A., & Bitterman, M. E. (1980). Some phenomena of associative learning in honeybees. *Journal of Comparative and Physiological Psychology, 94,* 878–885.

Couvillon, P. A., & Bitterman, M. E. (1982). Compound conditioning in honeybees. *Journal of Comparative and Physiological Psychology, 96,* 192–199.

Couvillon, P. A., & Bitterman, M. E. (1984). The overlearning-extinction effect and successive negative contrast in honeybees (*Apis mellifera*). *Journal of Comparative Psychology*, *98*, 100–109.

Couvillon, P. A., Lee, Y., & Bitterman, M. E. (1991). Learning in honeybees as a function of amount of reward: Rejection of the equal-asymptote assumption. *Animal Learning and Behavior*, *19*, 381–387.

Crane, E., & Graham, A. J. (1985). Bee hives of the ancient world. I. *Bee World*, *66*, 23–41.

Crothers, J. H. (1967). The biology of the shore crab *Carcinus maenas* (L.). 1. The background-anatomy, growth and life history. *Field Studies*, *2*, 407–434.

Crothers, J. H. (1968). The biology of the shore crab *Carcinus maenas* (L.). 2. The life of the adult crab. *Field Studies*, *3*, 579–614.

Crow, T. (1988). Cellular and molecular analysis of associative learning and memory in *Hermissenda*. *Trends in Neuroscience*, *11*, 136–142.

Crow, T. J. (1992). Associative learning in *Hermissenda*. In L. R. Squire (Ed.), *Encyclopedia of learning and memory* (pp. 293–298). New York: Macmillan.

Crow, T. J., & Alkon, D. L. (1978). Retention of an associative behavioral change in *Hermissenda*. *Science*, *201*, 1239–1241.

Cuadras, J., Vila, E., & Balasch, J. (1978). T-maze shock avoidance in the hermit crab *Dardanus arrosor*. *Revista Española de Fisiología*, *34*, 273–276.

Cunningham, P. N., & Hughes, R. N. (1984). Learning of predatory skills by shorecrabs *Carcinus maenas* feeding on mussels and dogwhelks. *Marine Ecology Progress Series*, *16*, 21–26.

Dalley, R., & Baily, H. (1981). A new apparatus used to record the locomotor rhythms of laboratory reared prawns and shrimps. *Marine Ecology Progress Series*, *4*, 229–234.

Daniel, P. C., & Derby, C. D. (1988). Behavioral olfactory discrimination of mixtures in the spiny lobster (*Panulirus argus*) based on a habituation paradigm. *Chemical Senses*, *13*, 385–395.

Darchen, R. (1964). Le substratum inné dans le comportement de *Blatella germanica* placée dans divers labyrinthes [Innate substrates and the behavior of *Blatella germanica* when placed in different mazes]. *Behaviour*, *22*, 245–281.

Darwin, C. H. (1936). *The origin of species by means of natural selection*. New York: Modern Library. (Original work published 1859)

Darwin, C. H. (1965). *The expression of the emotions in man and animals*. Chicago: University of Chicago Press. (Original work published 1872)

Dashevskii, B. A., Karas, A. Y., & Udalova, G. P. (1990). *Neuroscience and Behavioural Physiology*, *20*, 18–26.

Datta, L. G. (1962). Learning in the earthworm, *Lumbricus terrestris*. *American Journal of Psychology*, *75*, 531–553.

Datta, L. G., Milstein, S., & Bitterman, M. E. (1960). Habit reversal in the crab. *Journal of Comparative and Physiological Psychology*, *53*, 275–278.

Davis, M. (1970). Effects of interstimulus interval length and variability on startle-response habituation in the rat. *Journal of Comparative and Physiological Psychology*, *72*, 177–192.

Davis, M., & Wagner, A. R. (1968). Startle responsiveness after habituation to different intensities of tone. *Psychonomic Science*, *12*, 337–338.

Davis, W. J. (1986). Invertebrate model systems. In J. L. Martinez, Jr. & R. P. Kesner (Eds.), *Learning and memory: A biological view* (pp. 267–297). New York: Academic Press.

Debski, E., & Friesen, W. O. (1985). Habituation of swimming activity in the medicinal leech. *Journal of Experimental Biology*, *116*, 169–188.

DeCarlo, L. T., & Abramson, C. I. (1989). Time allocation in the carpenter ant (*Componotus herculeanus*). *Journal of Comparative Psychology*, *103*, 389–400.

DeJianne, D., McGuire, T. R., & Pruzan-Hotchkiss, A. (1985). Conditioned suppression of proboscis extension in *Drosophila melanogaster*. *Journal of Comparative Psychology*, *99*, 74–80.

Denny, M. R., & Ratner, S. C. (1970). *Comparative psychology: Research in animal behavior* (Rev. ed.). Homewood, IL: Dorsey Press.

Denti, A., Dimant, B., & Maldonado, H. (1988). Passive avoidance learning in the crab *Chasmagnathus granulatus*. *Physiology & Behavior*, *43*, 317–320.

Dethier, V. G. (1966). Insects and the concept of motivation. *Nebraska Symposium on Motivation*, *14*, 105–136.

Dethier, V. G. (1980). Food-aversion learning in two polyphagus caterpillars *Diacrisia virginica* and *Estigmene congrua*. *Physiological Entomology*, *5*, 321–325.

Dethier, V. G., Solomon, R. L., & Turner, L. H. (1965). Sensory input and central excitation and inhibition in the blowfly. *Journal of Comparative and Physiological Psychology*, *60*, 303–313.

Dickinson, A. (1980). *Contemporary animal learning theory*. Cambridge: Cambridge University Press.

Dinsmore, J. A. (1954). Punishment. I. The avoidance hypothesis. *Psychological Review*, *61*, 34–46.

Downey, P., & Jahan-Parwar, B. (1972). Cooling as reinforcing stimulus in *Aplysia*. *American Zoologist*, *12*, 507–512.

Drudge, O. W., & Platt, S. A. (1979). A versatile maze for learning and geotaxic selection in *Drosophila melanogaster*. *Behavior Research Methods & Instrumentation*, *11*, 503–506.

Drummond, H. (1985). Toward a standard ethogram: Do ethologists really want one? *Zeitschrift für Tierpsychologie*, *68*, 338–339.

Dudai, Y. (1977). Properties of learning and memory in *Drosophila melanogaster*. *Journal of Comparative Physiology*, *114*, 69–89.

Dudai, Y. (1992). Neurogenetic analysis of learning in *Drosophila*. In L. R. Squire (Ed.), *Encyclopedia of learning and memory* (pp. 307–310). New York: Macmillan.

Dunn, P. D. C., & Barnes, W. J. P. (1981a). Learning of leg position in the shore crab, *Carcinus maenas*. *Marine Behaviour and Physiology*, *8*, 67–82.

Dunn, P. D. C., & Barnes, W. J. P. (1981b). Neural correlates of leg learning in the shore crab, *Carcinus maenas*. *Marine Behaviour and Physiology*, *8*, 83–97.

Dyal, J. A. (1973). Behavior modification in annelids. In W. C. Corning, J. A. Dyal, & A. O. D. Willows (Eds.), *Invertebrate learning: Vol. 1. Protozoans through annelids* (pp. 225–290). New York: Plenum.

Dyal, J. A., & Corning, W. C. (1973). Invertebrate learning and behavioral taxonomies. In W. C. Corning, J. A. Dyal, & A. O. D. Willows (Eds.), *Invertebrate learning: Vol. 1* (pp. 1–48). New York: Plenum.

Dyal, J. A., & Hetherington, K. (1968). Habituation in the polychaete: *Hesperonoe adventor*. *Psychonomic Science*, *13*, 263–264.

Edwards, C. S., & Lofty, J. R. (1977). *Biology of earthworms* (2nd ed.). New York: Halsted Press.

Eisenstein, E. M. (1970). A comparison of activity and position response measures of avoidance learning in the cockroach, *P. americana*. *Brain Research*, *21*, 143–147.

Eisenstein, E. M., Brunder, D. G., & Blair, H. J. (1982). Habituation and sensitization in an aneural cell: Some comparative and theoretical considerations. *Neuroscience and Biobehavioral Reviews, 6,* 183–194.

Eisenstein, E. M., & Peretz, B. (1973). Comparative aspects of habituation in invertebrates. In H. V. S. Peeke & M. J. Herz (Eds.), *Habituation: Vol. 2. Physiological substrates* (pp. 1–34). New York: Academic Press.

Eisenstein, E. M., & Reep, R. L. (1985). Behavioral and cellular studies of learning and memory in insects. In G. A. Kerkut & L. I. Gilbert (Eds.), *Comprehensive insect physiology biochemistry and pharmacology: Vol. 9. Behaviour.* (pp. 514–540).

Emson, P., Walker, R. J., & Kerkut, G. A. (1971). Chemical changes in a molluscan ganglion associated with learning. *Comparative Biochemistry and Physiology, 40B,* 223–239.

Entingh, D., Dunn, A. J., Wilson, J. E., Glassman, E., Hogan, E., & Damstra, T. (1975). Biochemical approaches to the biological basis of memory. In M. S. Gazzaniga & C. Blakmore (Eds.), *Handbook of psychobiology* (pp. 201–238). New York: Academic Press.

Evans, S. M. (1963). Behaviour of the polychaete *Nereis* in T-mazes. *Animal Behaviour, 11,* 379–392.

Evans, S. M. (1966). Non-associative behavioral modifications in the polychaeta *Nereis diversicolor. Animal Behaviour, 14,* 107–112.

Evans, S. M. (1969a). Habituation of the withdrawal response in nereid polychaetes. 1. The habituation process in *Nereis diversicolor. Biological Bulletin, 137,* 95–104.

Evans, S. M. (1969b). Habituation of the withdrawal response in nereid polychaetes. 2. Rate of habituation in intact and decerebrate worms. *Biological Bulletin, 137,* 105–117.

Fantino, E., & Logan, C. A. (1979). *The experimental analysis of behavior: A biological perspective.* San Francisco: W. H. Freeman.

Farel, P. B., & Buerger, A. A. (1972). Instrumental conditioning of leg position in chronic spinal frog: Before and after sciatic section. *Brain Research, 47,* 345–351.

Farley, J. (1987a). Contingency learning and causal detection in *Hermissenda.* I. Behavior. *Behavioral Neuroscience, 101,* 13–27.

Farley, J. (1987b). Contingency learning and causal detection in *Hermissenda.* II. Cellular mechanisms. *Behavioral Neuroscience, 101,* pp. 28–56.

Farley, J., & Alkon, D. L. (1985). Cellular mechanisms of learning, memory, and information storage. *Annual Review of Psychology, 36,* 419–494.

Farley, J., & Alkon, D. L. (1987). Cellular analysis of gastropod learning. In A. H. Greenberg (Ed.), *Invertebrate models cell receptors and cell communication* (pp. 220–266). Basel: Karger.

Feinman, R. D., Abramson, C. I., & Forman, R. R. (1990). Classical conditioning in the crab. In K. Wiese & J. Tautz (Eds.), *Crustacean pioneer systems in neurobiology* (pp. 215–222). Berlin: Birkhäuser-Verlag.

Feinman, R. D., Korthals-Altes, H., Kingston, S., Abramson, C. I., & Forman, R. R. (1990). Lever-press conditioning in the crab. Green crabs perform well on fixed ratio schedules but can they count? *Biological Bulletin, 179,* 233 (abstract).

Feinman, R. D., Llinas, R. H., Abramson, C. I., & Forman, R. R. (1990). Electromyographic record of classical conditioning of eye withdrawal in the crab. *Biological Bulletin, 178,* 187–194.

Fernandez-Duque, E., Valeggia, C., & Maldonado, H. (1992). Multitrial inhibitory avoidance learning in the crab *Chasmagnathus. Behavioral and Neural Biology, 57,* 189–197.

Fielde, A. (1901). Further study of an ant. Proceedings of the *Academy of Natural Sciences of Philadelphia, 53*, 521–544.

Fine-Levy, J. B., Girardot, M. N., Derby, C. D., & Daniel, P. C. (1988). Differential associative conditioning and olfactory discrimination in the spiny lobster *Panulirus argus. Behavioral and Neural Biology, 49*, 315–331.

Florey, E. (1985). The zoological station at Naples and the neuron: Personalities and encounters in a unique institution. *Biological Bulletin, 168*, 137–152.

Florey, E. (1990). Crustacean neurobiology: History and perspectives. In K. Wiese, W. D. Krenz, J. Tautz, H. Reichert, & B. Mulloney (Eds.), *Frontiers in crustacean neurobiology* (pp. 4–32). Berlin: Birkhäuser-Verlag.

Forman, R. R. (1984). Leg position learning by an insect. I. A heat avoidance learning paradigm. *Journal of Neurobiology, 15*, 127–140.

Forman, R. R., & Zill, S. N. (1984). Leg position learning by an insect. II. Motor strategies underlying learned leg extension. *Journal of Neurobiology, 15*, 221–237.

Fraenkel, G. S., & Gunn, D. L. (1961). *The orientation of animals: Kineses, taxes and compass reactions.* New York: Dover Publications, Inc. (Original work published 1940)

France, R. L. (1985). Low pH avoidance by crayfish (*Orconectes virilis*): Evidence for sensory conditioning. *Canadian Journal of Zoology, 63*, 258–268.

Freud, S. Über den bau der nervenfasern und nervenzellen beim flusskrebs [The structure of nerve fibers and nerve cells in crayfish]. *Sitzungberichte Akademie der Wissenschafter in Wien. Mathematisch Naturwissenschaften Klasse. Kl. III Abt., 85*, 9–46.

Frings, H. (1944). The loci of olfactory end-organs in the honey bee. *Journal of Experimental Zoology, 97*, 123–134.

Frost, W. (1992). Habituation and sensitization in *Tritonia.* In L. R. Squire (Ed.), *Encyclopedia of learning and memory* (pp. 305–307). New York: Macmillan.

Fukushi, T. (1976). Classical conditioning to visual stimuli in the housefly, *Musca domestica. Journal of Insect Physiology, 22*, 361–364.

Fukushi, T. (1979). Properties of olfactory conditioning in the housefly, *Musca domestica. Journal of Insect Physiology, 25*, 155–159.

Fukushi, T. (1983). The role of learning on the finding of food in the searching behaviour of the housefly, *Musca domestica* (Diptera: Muscidae). *Entomologia Generalis, 8*, 241–250.

Fukushi, T. (1985). Visual learning in walking blowflies, *Lucilia cuprina. Comparative Physiology A, 157*, 771–778.

Fukushi, T. (1989). Learning and discrimination of colored papers in the walking blowfly, *Lucilia cuprina. Journal of Comparative Physiology (A): Sensory Neural and Behavioral Physiology, 166*, 57–64.

Fuller, J. L. (1960). Genetics and individual differences. In R. H. Waters, D. A. Rethlingshafer, & W. E. Caldwell (Eds.), *Principles of comparative psychology* (pp. 325–354). New York: McGraw-Hill.

Fuller, J. L., & Thompson, W. R. (1960). *Behavior genetics.* New York: Wiley.

Fuller, J. L., & Thompson, W. R. (1978). *Foundations of behavior genetics.* St. Louis: Mosby.

Gagné, R. M. (1965). *The conditions of learning.* New York: Holt, Rinehart and Winston.

Gallup, G. G., Jr. (1989). Editorial. *Journal of Comparative Psychology, 103*, 3.

Gandwere, S. K. (1960). The feeding and culturing of Orthoptera in the laboratory. *Entomology News, 71*, 7–45.

Gardiner, M. S. (1972). *The biology of invertebrates.* New York: McGraw-Hill.

Gardner, L. E. (1968). Retention and overhabituation of a dual-component response in *Lumbricus terrestris. Journal of Comparative and Physiological Psychology, 66*, 315–318.

Gardner, F. T., & Applewhite, P. B. (1970). Protein and RNA inhibitors and protozoan habituation. *Psychopharmacologia, 16*, 430–433.

Gelber, B. (1952). Investigations of the behavior of *Paramecium aurelia*. I. Modifications of behavior after training with reinforcement. *Journal of Comparative and Physiological Psychology, 45*, 58–65.

Gelperin, A. (1975). Rapid food aversion learning by a terrestrial mollusk. *Science, 189*, 567–570.

Gelperin, A. (1992). Associative learning in *Limax*. In L. R. Squire (Ed.), *Encyclopedia of learning and memory* (pp. 298–302). New York: Macmillan.

Getz, W. M., & Smith, K. B. (1987). Olfactory sensitivity and discrimination of mixtures in the honeybee *Alpis mellifera*. *Journal of Comparative Physiology A, 160*, 239–245.

Gillette, R. (1992). Associative learning in *Pleurobranchaea*. In L. R. Squire (Ed.), *Encyclopedia of learning and memory* (pp. 302–305). New York: Macmillan.

Goldman, C. R. (Ed.). (1983). *Freshwater crayfish*. Westport, CT: Avi Publishing.

Gonzalez, R. C., Longo, N., & Bitterman, M. E. (1961). Classical conditioning in the fish: Exploratory studies of partial reinforcement. *Journal of Comparative and Physiological Psychology, 54*, 452–456.

Goodman, L. J., & Fisher, R. C. (Eds.). *The behaviour and physiology of bees*. Slough, U.K.: C.A.B. International.

Gordon, D. M. (1985). Do we need more ethograms? *Zeitschrift für Tierpsychologie, 68*, 340–342.

Gormezano, I. (1984). The study of associative learning with CS-CR paradigms. In D. L. Alkon & J. Farley (Eds.), *Primary neural substrates of learning and behavioral change* (pp. 5–24). Cambridge: Cambridge University Press.

Gormezano, I., Kehoe, E. J., & Marshall, B. S. (1983). Twenty years of classical conditioning research with the rabbit. In J. M. Sprague & A. N. Epstein (Eds.), *Progress in psychobiology and physiological psychology: Vol. 10* (pp. 197–275), New York: Academic Press.

Gould, J. L. (1986). The biology of learning. *Annual Review of Psychology, 37*, 163–192.

Gould, J. L. (1991). The ecology of honeybee learning. In L. J. Goodman & R. C. Fisher (Eds.), *The behavior and physiology of bees* (pp. 306–322). London: C.A.B. International.

Graham, Jr., F. (1970). *Since silent spring*. Boston: Houghton Mifflin.

Graham, F. K. (1973). Habituation and dishabituation of responses innervated by the automatic nervous system. In H. V. S. Peeke & M. J. Herz (Eds.), *Habituation: Vol. 1. Behavioral studies* (pp. 163–218). New York: Academic Press.

Gray, P. H. (1973). Comparative psychology and ethology: A saga of twins reared apart. *Annals of the New York Academy of Sciences, 223*, 29–53.

Greenberg, S. M., Castellucci, V. F., Bayley, H., & Schwartz, J. H. (1987). A molecular mechanism for long-term sensitization in *Aplysia*. *Nature, 329*, 62–65.

Grossmann, K. E. (1973). Continuous, fixed-ratio and fixed-interval reinforcement in honey bees. *Journal of the Experimental Analysis of Behavior, 20*, 105–109.

Groves, P. M., & Thompson, R. F. (1970). Habituation: A dual-process theory. *Psychological Review, 77*, 419–450.

Hall, C. S. (1940). The genetics of behavior. In S. S. Stevens (Ed.), *Handbook of experimental psychology* (pp. 304–329). New York: Wiley.

Hansell, M. (1985). Ethology. In G. A. Kerkut & L. I. Gilbert (Eds.), *Comprehensive insect physiology biochemistry and pharmacology: Vol 9. Behavior* (pp. 1–93). Oxford: Pergamon.

Haralson, J. V., Groff, C. I., & Haralson, S. J. (1975). Classical conditioning in the sea anemone, *Cribrina xanthogrammica*. *Physiology & Behavior, 15*, 455–460.

Haralson, S. J., & Haralson, J. V. (1988). Habituation in the sea anemone (*Anthopleura elegantissima*): Spatial discrimination. *International Journal of Comparative Psychology, 1*, 245–253.

Harless, M. D. (1967). Successive reversals of a position response in isopods. *Psychonomic Science, 9*, 123–124.

Harré, R., & Lamb, R. (Ed.). (1986). *The dictionary of ethology and animal learning.* Cambridge, MA: MIT Press.

Harris, C. L. (1991). An improved Horridge procedure for studying leg-position learning in cockroaches. *Physiology & Behavior, 49*, 543–548.

Harris, J. D. (1943). Habituatory response decrement in intact organisms. *Psychological Bulletin, 40*, 385–422.

Hawkins, R. D. (1991). Cell biological studies of conditioning in *Aplysia*. In J. Madden IV (Ed.), *Neurobiology of learning, emotion and affect* (pp. 3–28). New York: Raven Press.

Hawkins, R. D., & Bower, G. H. (Eds.). (1989). Computational models of learning in simple neural systems. *Psychology of learning and motivation: Vol 23*. New York: Academic Press.

Hawkins, R. D., & Bruner, J. (1981). Activity of excitor and inhibitor claw motor neurons during habituation and dishabituation of the crayfish defense response. *Journal of Experimental Biology, 91*, 145–164.

Hawkins, R. D., Carew, T. J., & Kandel, E. R. (1986). Effects of interstimulus interval and contingency on classical conditioning of the *Aplysia* siphon withdrawal reflex. *Journal of Neuroscience, 6*, 1695–1701.

Hawkins, R. D., & Kandel, E. R. (1984). Is there a cell-biological alphabet for simple forms of learning? *Psychological Review, 91*, 375–391.

Hay, D. A., & Crossley, S. A. (1977). The design of mazes to study *Drosophila* behavior. *Behavior Genetics, 7*, 389–402.

Hearst, E. (1988). Fundamentals of learning and conditioning. In R. C. Atkinson, R. J. Herrnstein, G. Lindzey, & R. D. Luce (Eds.), *Steven's handbook of experimental psychology: Vol. 2. Learning and cognition* (pp. 3–109). New York: Wiley.

Henderson, T. B., & Strong, P. N. (1972). Classical conditioning in the leech *Macrobdella ditetra* as a function of CS and UCS intensity. *Conditioned Reflex, 7*, 210–215.

Hennessey, T. M., Rucker, W. B., & McDiarmid, C. G. (1979). Classical conditioning in paramecia. *Animal Learning and Behavior, 7*, 417–423.

Herz, M. J., Peeke, H. V. S., & Wyers, E. J. (1964). Temperature and conditioning in the earthworm *Lumbricus terrestris*. *Animal Behaviour, 12*, 502–507.

Hewitt, J. K., Fulker, D. W., & Hewitt, C. A. (1983). Genetic architecture of olfactory discriminative avoidance conditioning in *Drosophila melanogaster*. *Journal of Comparative Psychology, 97*, 52–58.

Hilgard, E. R. (1951). Methods and procedures in the study of learning. In S. S. Stevens (Ed.), *Handbook of experimental psychology* (pp. 517–567). New York: Wiley.

Hilgard, E. R., & Marquis, D. G. (1940). *Conditioning and learning.* New York: Appleton-Century-Crofts.

Hill, W. F. (1977). *Learning: A survey of psychological interpretations* (3rd ed.). New York: Crowell.

Hinde, R. A. (1970). Behavioural habituation. In G. Horn & R. A. Hinde (Eds.), *Short-term changes in neural activity and behaviour* (pp. 3–40). New York: Cambridge University Press.

Hinde, R. A. (1982). *Ethology.* Oxford: Oxford University Press.

Hirsch, J., & Erlenmeyer-Kimling, L. (1967). Behavioral genetics. In H. Helson & W. Bevan (Eds.), *Contemporary approaches to psychology* (pp. 91–137). Princeton, NJ: D. Van Nostrand.

Hirsch, J., & Holliday, M. (1988). A fundamental distinction in the analysis and interpretation of behavior. *Journal of Comparative Psychology, 102,* 372–377.

Hoagland, H. (1931). A study of physiology of learning in ants. *Journal of General Psychology, 5,* 21–41.

Hodgkin, A. L., & Huxley, A. F. (1945). Resting and action potentials in single nerve fibers. *Journal of Physiology, 104,* 176–195.

Hodos, W., & Campbell, C. B. G. (1990). Evolutionary scales and comparative studies of animal cognition. In R. P. Kesner & D. S. Olton (Eds.), *Neurobiology of comparative cognition* (pp. 1–20). Hillsdale, NJ: Lawrence Erlbaum Associates.

Hölldobler, B., & Wilson, E. O. (1990). *The ants.* Cambridge: Belknapp Press.

Horridge, G. A. (1962). Learning of leg position by headless insects. *Nature, 193,* 697–698.

Hoyle, G. (1976). Learning of leg position by the ghost crab *Ocyppode ceratophthalma. Behavioral Biology, 18,* 147–163.

Hoyle, G. (1980). Learning using natural reinforcements in insect preparations that permit cellular neuronal analysis. *Journal of Neurobiology, 11,* 323–354.

Hoyle, G. (1988). Behavior in the light of identified neurons. In A. C. Catania & S. Harnad (Eds.), *The selection of behavior: The operant behaviorism of B. F. Skinner: Comments and consequences* (pp. 434–436). Cambridge: Cambridge University Press.

Hull, C. L. (1943). *Principles of behavior.* New York: Appleton-Century-Crofts.

Hull, C. L. (1945). The place of individual species differences in a natural-science theory of behavior. *Psychological Review, 52,* 55–60.

Humphrey, G. (1930). Le Chatelier's rule, and the problem of habituation and dishabituation in *Helix albolabris. Psychologische Forschung, 13,* 113–127.

Huxley, T. H. (1973). *The crayfish: An introduction to the study of zoology.* Cambridge, MA: The MIT Press. (Original work published 1880)

Hyman, L. H. (1940–1967). *The invertebrates,* (Vols. 1–6). New York: McGraw-Hill.

Inozemtsev, A. N. (1990). Changes in behavioral responses of nereids to vibration after unconditioned stimulation. *Neuroscience and Behavioural Physiology, 20,* 185–187.

Ishida, M., Couvillon, P. A., & Bitterman, M. E. (1992). Acquisition and extinction of a shuttling response in honeybees (*Apis mellifera*) as a function of the probability of reward. *Journal of Comparative Psychology, 106,* 362–369.

Jacobson, A. L., Fried, C., & Horowitz, S. D. (1967). Classical conditioning, pseudoconditioning, or sensitization in the planarian. *Journal of Comparative and Physiological Psychology, 64,* 73–79.

Jaynes, J. (1969). The historical origins of "ethology" and "comparative psychology." *Animal Behaviour, 17,* 601–606.

Jennings, H. S. (1976). *Behavior of the lower organisms.* Bloomington: Indiana University Press. (Original work published 1906)

Jones, O. T., Lower, R. A., & Howse, P. E. (1981). Responses of male Mediterranean fruit flies, (*Ceratitis capitata*), to timed lure in a wind tunnel of novel design. *Physiological Entomology, 6,* 175–181.

Joravsky, D. (1989). *Russian psychology: A critical history.* Cambridge: Basil Blackwell.

Kaiser, F. (1954). Beiträge zur bewegungsphysiologie der hirudeen [Contributions to the physiology of motion of the leeches]. *Zoologische Jahrbuecher, 65,* 59–90.

Kandel, E. R. (1970). Nerve cells and behavior. *Scientific American, 223,* 57–70.

Kandel, E. R. (1979). Cellular aspects of learning. In M. A. B. Brazier (Ed.), *Brain mech-*

anisms in memory and learning: From the single neuron to man (pp. 3–16). New York: Raven Press.

Kandel, E. R., & Tauc, L. (1965). Mechanisms of heterosynaptic facilitation in the giant cell of the abdominal ganglion of *Aplysia depilans. Journal of Physiology, 181*, 28–47.

Karas, A. Y. (1962). Feeding conditioned reflexes from visual, tactile, and static receptors in the crab *Carcinus maenas. Nauchniye Doklady Vysshey Shkoly Biologicheskiye Nauki, 2*, 748–755.

Kennedy, D. (1967). Small systems of nerve cells. *Scientific American, 216*, 44–52.

Keshavamurthy, P., & Krishnamoorthy, R. V. (1977). A circadian rhythm in the electrode-avoidance behavior of the earthworm *Megascolex mauritii, Pheretima elongata*, and *Perionyx excavatus. Behavioral Biology, 20*, 17–24.

Kety, S. S. (1968). A biologist examines the mind and behavior. In W. C. Corning & M. Balaban (Eds.), *The mind: Biological approaches to its function* (pp. 283–310). New York: Wiley Interscience.

Kimble, G. A. (1961). *Hilgard and Marquis' conditioning and learning.* New York: Appleton-Century-Crofts.

Kimmel, H. D., & Yaremko, R. M. (1966). Effect of partial reinforcement on acquisition and extinction of classical conditioning in the planarian. *Journal of Comparative and Physiological Psychology, 61*, 299–301.

Kirchner, W. H., Dreller, C., & Towne, W. F. (1991). Hearing in honeybees: Operant conditioning and spontaneous reactions to airborne sound. *Journal of Comparative Physiology A, 168*, 85–89.

Kirk, W. E., & Thompson, R. W. (1967). Effects of light, shock, and goal box conditions on runway performance of the earthworm, *Lumbricus terrestris. Psychological Record, 17*, 49–54.

Kirkpatrick, T., Schneider, C. W., & Pavloski, R. (1991). A computerized infrared monitor for following movement in aquatic animals. *Behavior Research Methods and Instruments and Computers, 23*, 16–22.

Klopf, A. H. (1988). A neuronal model of classical conditioning. *Psychobiology, 16*, 85–125.

Krasne, F. B. (1969). Excitation and habituation of the crayfish escape reflex: The depolarizing response in lateral giant fibers of the isolated abdomen. *Journal of Experimental Biology, 50*, 29–46.

Krasne, F. B. (1973). Learning in crustacea. In W. C. Corning, J. A. Dyal, & A. O. D. Willows (Eds.), *Invertebrate learning· Vol. 2. Arthropods and gastropod mollusks* (pp. 49–130). New York: Plenum.

Krasne, F. B. (1984). Physiological analysis of learning in invertebrates. In F. Reinosa-Suarez & C. Ajmone-Marsan (Eds.), *Cortical integration* (pp. 53–76). New York: Raven Press.

Krasne, F. B. (1992). Nonassociative learning in crayfish. In L. R. Squire (Ed.), *Encyclopedia of learning and memory* (pp. 310–311). New York: Macmillan.

Kuenzer, P. P. (1958). Verhaltenphysiologische untersuchungen über das zucken des regenwürms [Behavioral physiological research on the retraction reflex of the rainworm]. *Zeitschrift für Tierpsychologie, 15*, 31–49.

Lahue, R., & Corning, W. C. (1971). Habituation in *Limulus* abdominal ganglia. *Biological Bulletin, 140*, 427–439.

Land, M. F. (1971). Orientation by jumping spiders in the absence of visual feedback. *Journal of Experimental Biology, 54*, 119–139.

Laybourn-Parry, J. (1984). *A functional biology of free-living protozoa.* Berkeley, CA: University of California Press.

LeBourg, E. (1983). Aging and habituation of the tarsal response in *Drosophila melanogaster*. *Gerontology, 29*, 388–393.

Lederhendler, I. I., Gart, S., & Alkon, D. L. (1986). Classical conditioning of *Hermissenda*: Origin of a new reponse. *Journal of Neuroscience, 6*, 1325–1331.

Lee, J. J., Hunter, S. H., & Bovee, E. C. (1985). *Illustrated guide to the protozoa*. Lawrence, KS: Society of Protozoologists.

Lee, V. L. (1988). *Beyond behaviorism*. Hillsdale, NJ: Lawrence Erlbaum Associates.

Lee, Y., & Bitterman, M. E. (1990a). Learning in honeybees (*Apis mellifera*) as a function of amount of reward: Acquisition measures. *Journal of Comparative Psychology, 104*, 152–158.

Lee, Y., & Bitterman, M. E. (1990b). Learning in honeybees (*Apis mellifera*) as a function of amount of reward: Control of delay. *Animal Learning & Behavior, 18*, 377–386.

Lee, S. K., & Bitterman, M. E. (1992). Learning in honeybees (*Apis mellifera*) as a function of sucrose concentration. *Journal of Comparative and Physiological Psychology, 106*, 29–36.

Leeming, F. C. (1985). Free-response escape but not avoidance learning in houseflies (*Musca domestica*). *Psychological Record, 35*, 513–523.

Leeming, F. C., & Little, G. L. (1977). Learning in houseflies (*Musca domestica*). *Journal of Comparative and Physiological Psychology, 91*, 260–269.

Leibrecht, B. C. (1972). Habituation, 1940–1970: Bibliography and key word index. *Psychonomic Monograph Supplements, 4*, 189–217.

Leibrecht, B. C. (1974). Habituation: Supplemental bibliography. *Physiological Psychology, 2*, 1–19.

Lejeune, H., & Richelle, M. (1990). Timing behavior and development: Comments on some animal and human data. *International Journal of Comparative Psychology, 4*, 111–135.

Lejeune, H., & Wearden, J. H. (1991). The comparative psychology of fixed-interval responding: Some quantitative analysis. *Learning and Motivation, 22*, 84–111.

Lenhoff, H. M. (Ed.). (1983). *Hydra: Research methods*. New York: Plenum.

Lenhoff, H. M., & Lenhoff, S. G. (1988). Trembley's polyps. *Scientific American, 258*, 108–113.

Lennartz, R. C., & Weinberger, N. M. (1992). Analysis of response systems in Pavlovian conditioning reveals rapidly versus slowly acquired conditioned responses: Support for two factors, implications for behavior and neurobiology. *Psychobiology, 20*, 93–119.

Leonard, J. L., & Lukowiak, K. (1985). The standard ethogram: A two-edged sword? *Zeitschrift für Tierpsychologie, 68*, 335–337.

Leonard, J. L., & Lukowiak, K. (1986). The behavior of *Aplysia californica* Cooper (Gastropoda; opisthobranchia): I. Ethogram. *Behaviour, 98*, 320–360.

Lester, D. (1973). *Comparative psychology: Phyletic differences in behavior*. New York: Alfred Publishing Company.

Levine, D. S. (1983). Neural population modeling and psychology: A review. *Mathematical Biosciences, 66*, 1–86.

Levine, D. S. (1989). Neural network principles for theoretical psychology. *Behavior Research Methods, Instruments, & Computers, 21*, 213–224.

Lillie, F. R. (1988). *The Woods Hole Marine Biological Laboratory*. Chicago: University of Chicago Press. (Reprinted in Vol. 174, Supplement 1 of the *Biological Bulletin*.) (Original work published 1944)

Lloyd, D. (1986). The limits of cognitive liberalism. *Behaviorism, 14*, 1–14.

Lockard, R. B. (1968). The albino rat: A defensible choice or a bad habit? *American Psychologist, 23*, 734–742.

Lockery, S. R., Rawlins, J. N. P., & Gray, J. A. (1985). Habituation of the shortening reflex in the medicinal leech. *Behavioral Neuroscience, 99*, 333–341.

Loeb, J. (1973). *Forced movements, tropisms, and animal conduct.* New York: Dover Publications, Inc. (Original work published 1918)

Logan, C., & Beck, H. (1978). Long term retention of habituation in the absence of a central nervous system. *Journal of Comparative and Physiological Psychology, 92*, 928–934.

Logan, F. A. (1960). *Incentive: How the conditions of reinforcement affect the performance of rats.* New Haven: Yale University Press.

Logunov, D. B. (1981). Conditioned reflex to time in *Helix locorum. Neuroscience and Behavioural Physiology, 11*, 234–240.

Longo, N. (1964). Probability learning and habit reversal in the cockroach. *American Journal of Psychology, 77*, 29–41.

Longo, N. (1970). A runway for the cockroach. *Behavior Research Methods & Instrumentation, 2*, 118–119.

Longo, N., Milstein, S., & Bitterman, M. E. (1962). Classical conditioning in the pigeon: Exploratory studies of partial reinforcement. *Journal of Comparative and Physiological Psychology, 55*, 983–986.

Lozada, M., Romano, A., & Maldonado, H. (1988). Effect of morphine and naloxone on a defensive response of the crab *Chasmagnathus granulatus. Pharmacology Biochemistry & Behavior, 30*, 635–640.

Lubow, R. E. (1973). Latent inhibition. *Psychological Bulletin, 79*, 398–407.

Lukowiak, K., & Sahley, C. (1981). The in vitro classical conditioning of the gill withdrawal reflex of *Aplysia californica. Science, 212*, 1516–1518.

Lutz, F. E., Welch, P. S., Galtsoff, P. S., & Needham, J. G. (Eds.). (1959). *Culture methods for invertebrate animals.* New York: Dover. (Original work published 1937)

Lutz, P. E. (1986). *Invertebrate zoology.* Menlo Park, CA: Benjamin/Cummings.

Mackintosh, N. J. (1974). *The psychology of animal learning.* New York: Academic Press.

Mackintosh, N. J. (1983). *Conditioning and associative learning.* Oxford: Oxford University Press.

Macphail, E. M. (1987). The comparative psychology of intelligence. *Behavioral and Brain Sciences, 10*, 645–695.

Maes, F. W., & Bijpost, S. C. A. (1979). Classical conditioning reveals discrimination of salt taste quality in the blowfly. *Journal of Comparative Physiology, 133A*, 53–62.

Maier, N. R. F., & Schneirla, T. C. (1942). Mechanisms in conditioning. *Psychological Review, 49*, 117–134.

Maier, N. R. F., & Schneirla, T. C. (1964). *Principles of animal psychology* (rev. ed.). New York: Dover. (Original work published 1935)

Makous, W. L. (1969). Conditioning in the horseshoe crab. *Psychonomic Science, 14*, 4–6.

Maldonado, H. (1970). An automatic recorder to register the praying mantis' response to the presence of a moving object. *Physiology & Behavior, 5*, 1337–1340.

Mamood, A. N., & Waller, G. D. (1990). Recovery of learning responses by honeybees following a sublethal exposure to permethrin. *Physiological Entomology, 15*, 55–60.

Mantel, L. H. (Organizer). 1992. The compleat crab. [Special issue; symposium presented at the Annual Meeting of the American Society of Zoologists, December 27–30, 1990, San Antonio, TX]. *American Zoologist, 32*(3).

Marian, R. W., & Abramson, C. I. (1982). Earthworm behavior in a modified running wheel. *Journal of Mind and Behavior, 3*, 67–74.

Mariath, H. A. (1985). Operant conditioning in *Drosophila melanogaster* wild-type and learning mutants with defects in the cyclic AMP metabolism. *Journal of Insect Physiology, 31*, 779–787.

Martin, P., & Bateson, P. (1986). *Measuring behaviour: An introductory guide*. Cambridge: Cambridge University Press.

Martinsen, D. L., & Kimeldorf, D. J. (1972). Conditioned spatial avoidance behavior of ants induced by x-rays. *Psychological Record, 22*, 225–232.

May, M. L., & Hoy, R. R. (1991). Habituation of the ultrasound-induced acoustic startle response in flying crickets. *Journal of Experimental Biology, 159*, 489–499.

Mayr, E. (1982). *The growth of biological thought: Diversity, evolution, and inheritance*. Cambridge, MA: Belknap Press.

Mazur, J. E. (1994). *Learning and behavior* (3rd ed.). Englewood Cliffs, NJ: Prentice Hall.

McClearn, G. E., & Foch, T. T. (1988). Behavioral genetics. In R. C. Atkinson, R. J. Herrnstein, G. Lindzey, & R. D. Luce (Eds.), *Steven's handbook of experimental psychology: Vol. 1. Perception and motivation* (2nd ed.; pp. 677–764). New York: Wiley.

McConnell, J. V. (1962). Memory transfer through cannibalism in planarians. *Journal of Neuropsychiatry, 3*, 42–48.

McConnell, J. V. (Ed.). (1967a). *A manual of psychological experiments on planarians*. Ann Arbor: Journal of Biological Psychology.

McConnell, J. V. (1967b). Specific factors influencing planarian behavior. In W. C. Corning & S. C. Ratner (Eds.), *Chemistry of learning* (pp. 217–233). New York: Plenum.

McConnell, J. V., Jacobson, A. L., & Kimble, D. P. (1959). The effects of regeneration upon retention of a conditioned response in the planarian. *Journal of Comparative and Physiological Psychology, 52*, 1–5.

McConnell, J. V., & Shelby, J. M. (1970). Memory transfer experiments in invertebrates. In G. Unger (Ed.), *Molecular mechanisms in memory and learning* (pp. 71–101). New York: Plenum.

McDaniel, J. W. (1969). Reversal learning in the sow bug. *Psychonomic Science, 16*, 261–262.

McGaugh, J. L., Weinberger, N. M., & Whalen, R. E. (1967). *Readings from Scientific American: Psychbiology*. San Francisco: W. H. Freeman.

McGuire, T. R. (1984). Learning in three species of diptera: The blow fly *Phormia regina*, the fruit fly *Drosophila melanogaster*, and the house fly *Musca domestica*. *Behavior Genetics, 14*, 479–526.

McManus, F. E., & Wyers, E. J. (1979). Olfaction and selection association in the earthworm, *Lumbricus terrestris*. *Behavioral and Neurobiology, 25*, 39–57.

Menzel, R. (1992). Associative learning in bees. In L. R. Squire (Ed.), *Encyclopedia of learning and memory* (pp. 289–293). New York: Macmillan.

Menzel, R., & Erber, J. (1972). The influence of the quantity of reward on the learning performance in honeybees. *Behaviour, 41*, 27–42.

Menzel, R., Erber, J., & Masuhr, T. (1974). Learning and memory in the honeybee. In L. B. Brown (Ed.), *Experimental analysis of insect behavior* (pp. 195–217). Heidelberg: Springer-Verlag.

Menzel, R., Hammer, M., Braun, G., Mauelshagen, J., & Sugawa, M. (1991). Neurobiology of learning and memory in honeybees. In L. J. Goodman & R. C. Fisher (Eds.), *The behaviour and physiology of bees* (pp. 323–353). Slough, U.K.: C.A.B. International.

Miall, R. C., & Hereward, C. J. (1988). A simple miniature capacitive position transducer. *Journal of Experimental Biology, 138*, 541–544.

Mikhailoff, S. (1922). Expérience réflexologique: Expérience nouvelles sur *Pagurus stria-*

tus. (Reflexological experiences: New experiences of *Pagurus striatus*). *Bulletin de L'Institut Océanographique*. Monaco, No. 418.

Mikhailoff, S. (1923). Expérience réflexologique: Expériences nouvelles sur *Pagurus striatus, Leander xiphiaas* et *treillianus* [Reflexological experiences: New experiences of *Pagurus striatus, Leander xiphiaas*, and *treillianus*]. *Bulletin de L'Institut Océanographique*. Monaco, No. 422.

Miller, J. R., & Roelofs, W. L. (1978). Sustained-flight tunnel for measuring insect responses to windborne sex pheromones. *Journal of Chemical Ecology, 4,* 187–198.

Miller, T. A. (1979). *Insect neurophysiological techniques*. New York: Springer-Verlag.

Miller, T., Bruner, L. J., & Fukuto, T. R. (1971). The effect of light, temperature, and DDT poisoning in housefly locomotion and flight muscle activity. *Pesticide Biochemistry and Physiology, 1,* 483–491.

Mims, F. M., III. (1986). *Engineer's mini-notebook: Optoelectronic circuits*. Fort Worth, TX: Radio Shack.

Minami, H., & Dallenbach, K. M. (1946). The effect of activity upon learning and retention in the cockroach. *American Journal of Psychology, 59,* 1–58.

Moore, D., & Rankin, M. A. (1983). Diurnal changes in the accuracy of the honeybee foraging rhythm. *Biological Bulletin, 164,* 471–482.

Morgan, R. F. (1981). Learning in the submerged *Formica rufa*. *Psychological Reports, 49,* 63–69.

Morgan, R. F., Ratner, S. C., & Denny, M. R. (1965). Response of earthworms to light as measured by the GSR. *Psychonomic Science, 3,* 27–28.

Mpitsos, G., & Collins, S. (1975). Learning: Rapid aversive conditioning in the gastropod mollusk *Pleurobranchaea*. *Science, 188,* 954–957.

Mpitsos, G. J., & Cohan, C. S. (1986). Differential conditioning in the mollusc *Pleurobranchaea*. *Journal of Neurobiology, 17,* 487–497.

Mpitsos, G. J., & Davis, W. J. (1973). Learning: Classical and avoidance conditioning in the mollusk *Pleurobranchaea*. *Science, 180,* 317–320.

Neher, E., & Sakmann, B. (1992). The patch clamp technique. *Scientific American, 266,* 44–51.

Nelson, M. C. (1971). Classical conditioning in the blowfly (*Phormia regina*): Associative and excitatory factors. *Journal of Comparative and Physiological Psychology, 77,* 353–368.

Nicholls, J. G., & Van Essen, D. (1974). The nervous system of the leech. *Scientific American, 230,* 38–48.

Oakley, B., & Schafer, R. (1978). *Experimental neurobiology: A laboratory manual*. Ann Arbor: The University of Michigan Press.

Offutt, G. C. (1970). Acoustic stimulus perception by the American lobster *Homarus americanus* (Decapoda). *Experientia, 26,* 1276–1278.

Olivo, R. F., & Thompson, M. C. (1988). Monitoring animals' movement using digitized video images. *Behavior Research Methods Instruments, and Computers, 20,* 485–490.

Olson, G. S., & Strandberg, R. (1979). Instrumental conditioning in crayfish: Lever pulling for food. *Society for Neuroscience Abstracts, 5,* 257.

Osborn, D., Blair, H. J., Thomas, J., & Eisenstein, E. M. (1973). The effects of vibratory and electrical stimulation on habituation in the ciliated protozoan, *Spirostomum ambiguum*. *Behavioral Biology, 8,* 655–664.

Pak, K., & Harris, C. L. (1975). Evidence for a molecular code for learned shock-avoidance in cockroaches. *Comparative Biochemistry and Physiology, 52A,* 141–144.

Pantaleâo, G., & Morato, S. (1989). A new method for the study of feeding behavior in the fly *Ceratitis capitata*. *Bulletin of the Psychonomic Society, 27,* 274–276.

Papaj, D. R., & Prokopy, R. J. (1989). Ecological and evolutionary aspects of learning in phytophagous insects. *Annual Review of Entomology*, *34*, 315–350.

Papini, M. R., & Bitterman, M. E. (1990). The role of contingency in classical conditioning. *Psychological Review*, *97*, 396–403.

Pavlov, I. P. (1927). *Conditioned reflexes*. Oxford: Oxford University Press.

Pavlov, I. P. (1951). How does the clam open its folds? *Complete Works* (Vol. 1; pp. 466–493). (Original work published 1885)

Pearse, V., Pearse, J., Buchsbaum, M., & Buchsbaum, R. (1987). *Living invertebrates*. Palo Alto, CA: Blackwell Scientific Publications.

Peckham, G. W., & Peckham, E. G. (1887). Some observations on the mental power of spiders. *Journal of Morphology*, *1*, 383–419.

Peeke, H. V. S., Herz, M. J., & Wyers, E. J. (1965). Amount of training, intermittent reinforcement and resistance to extinction of the conditioned withdrawal responses in the earthworm (*Lumbricus terrestris*). *Animal Behaviour*, *13*, 566–570.

Peretz, B. (1970). Habituation and dishabituation in the absence of a central nervous system. *Science*, *9*, 379–381.

Peretz, B., & Lukowiak, K. (1975). Age-dependent CNS control of the habituating gill withdrawal reflex and correlated activity in identified neurons in *Aplysia*. *Journal of Comparative Physiology*, *103*, 1–17.

Pessotti, I. (1972). Discrimination with light stimuli and a lever-pressing response in *Melipona rufiventris*. *Journal of Apicultural Research*, *11*, 89–93.

Pessotti, I., & Lignelli-Otero, V. R. (1981). Aprendizagem em abelhas. IV. Punicao e resistencia a extincao [Understanding in bees. IV. Punishment and resistance to extinction]. *Revista Brasileira de Brasileira de Biologia*, *41*, 673–680.

Pinsker, H., Kupfermann, I., Castellucci, V., & Kandel, E. (1970). Habituation and dishabituation of the gill-withdrawal reflex in *Aplysia*. *Science*, *167*, 1740–1742. ·

Platt, S. A., Holliday, M., & Drudge, O. W. (1980). Discrimination learning of an instrumental response in individual *Drosophila melanogaster*. *Journal of Experimental Psychology: Animal Behavior Processes*, *6*, 301–311.

Plomin, R., Defries, J. C., & McClearn, G. E. (1980). *Behavioral genetics: A primer*. San Francisco: W. H. Freeman.

Preston, R. J., & Lee, R. M. (1973). Feeding behavior in *Aplysia californica*: Role of chemical and tactile stimuli. *Journal of Comparative and Physiological Psychology*, *82*, 368–381.

Pribram, K. H., & Broadbent, D. E. (Eds.). (1970). *Biology of memory*. New York: Academic Press.

Pritchatt, D. (1968). Avoidance of electric shock by the cockroach *Periplaneta americana*. *Animal Behaviour*, *16*, 178–185.

Pritchatt, D. (1970). Further studies on the avoidance behavior of *Periplaneta americana*. *Animal Behaviour*, *18*, 485–492.

Prosser, C. L. (1934). Action potentials in the nervous system of the crayfish: II. Responses to illumination of the eye and caudal ganglion. *Journal of Cellular and Comparative Physiology*, *4*, 363–377.

Provenzano, A. J. (Ed.). (1985). *The biology of crustacea: Vol. 10. Economic aspects: Fisheries and culture*. New York: Academic Press.

Punzo, F. (1983). Localization of brain function and neuralchemical correlates of learning in the mud crab, *Eurypanopeus depressus* (Decapod). *Comparative Biochemistry and Physiology (A)*, *75*, 299–305.

Quinn, W. G., Harris, W. A., & Benzer, S. (1974). Conditioned behavior in *Drosophila melanogaster*. *Proceedings of the National Academy of Science*, *71*, 708–712.

Rachlin, H. (1970). *Introduction to modern behaviorism*. San Francisco: W. H. Freeman.

Ragland, R. S., & Ragland, J. B. (1965). Planaria: Interspecific transfer of a conditionability factor through cannibalism. *Psychonomic Science, 3,* 117–118.

Rakitin, A., Tomsic, D., & Maldonado, H. (1991). Habituation and sensitization to an electric shock in the crab *Chasmagnathus.* Effect of background illumination. *Physiology & Behavior, 50,* 477–487.

Rankin, C. H., Beck, C. D. O., & Chiba, C. M. (1990). *Caenorhabitis elegans:* A new model system for learning and memory. *Behavioural Brain Research, 37,* 89–92.

Rankin, C. H., & Broster, B. S. (1992). Factors affecting habituation and recovery from habituation in the nematode *Caenorhabitis elegans. Behavioral Neuroscience, 106,* 239–249.

Rankin, C. H., & Carew, T. J. (1988). Dishabituation and sensitization emerge as separate processes during development in *Aplysia. Journal of Neuroscience, 8,* 197–211.

Ratner, S. C. (1962). Conditioning of decerebrate worms, *Lumbricus terrestris. Journal of Comparative and Physiological Psychology, 55,* 174–177.

Ratner, S. C. (1965). Research and theory on conditioning of annelids. In D. Davenport & W. H. Thorpe (Eds.), *Learning and associated phenomena in invertebrates* (pp. 101–108). *Animal Behavior,* Supplement 1.

Ratner, S. C. (1967). Annelids and learning: A critical review. In W. C. Corning & S. C. Ratner (Eds.), *Chemistry of learning* (pp. 391–406). New York: Plenum.

Ratner, S. C. (1968). Reliability of indexes of worm learning. *Psychological Reports, 22,* 130.

Ratner, S. C. (1970). Habituation: Research and theory. In J. H. Reynierse (Ed.), *Current issues in animal learning* (pp. 55–84). Lincoln: University of Nebraska Press.

Ratner, S. C. (1972). Habituation and retention of habituation in the leech (*Macrobdella decora*). *Journal of Comparative and Physiological Psychology, 81,* 115–121.

Ratner, S. C. (1980). The comparative method. In M. R. Denny (Ed.), *Comparative psychology: An evolutionary analysis of animal behavior* (pp. 152–167). New York: Wiley.

Ratner, S. C., & Boice, R. (1971). Behavioral characteristics and functions of pheromones of earthworms. *Psychological Record, 21,* 363–371.

Ratner, S. C., & Gardner, L. E. (1968). Variables affecting responses of earthworms to light. *Journal of Comparative and Physiological Psychology, 66,* 239–243.

Ratner, S. C., & Miller, K. R. (1959a). Classical conditioning in earthworms, *Lumbricus terrestris. Journal of Comparative and Physiological Psychology, 52,* 102–105.

Ratner, S. C., & Miller, K. R. (1959b). Effects of spacing of training and ganglia removal on conditioning in earthworms. *Journal of Comparative and Physiological Psychology, 52,* 667–672.

Ratner, S. C., & Stein, D. G. (1965). Responses of worms to light as a function of intertrial interval and ganglion removal. *Journal of Comparative and Physiological Psychology, 59,* 301–304.

Ray, A. J. (1968). Instrumental light avoidance by the earthworm. *Communications in Behavioral Biology, 1,* 205–208.

Razran, G. (1971). *Mind in evolution: An east-west synthesis of learned behavior and cognition.* Boston: Houghton Mifflin.

Real, L. A. (1991). Animal choice behavior and the evolution of cognitive architecture. *Science, 253,* 980–986.

Reep, R. L., Eisenstein, E. M., & Tweedle, C. D. (1980). Neuronal pathways involved in transfer of information related to leg position in the cockroach, *P. americana. Physiology & Behavior, 24,* 501–513.

Rescorla, R. A. (1980). Simultaneous and successive associations in sensory preconditioning. *Journal of Experimental Psychology: Animal Behavior Processes, 6,* 207–216.

Rescorla, R. A., & Cunningham, C. L. (1978). Within-compound flavor associations. *Journal of Experimental Psychology: Animal Behavior Processes, 4*, 267–275.

Ressler, R., Cialdini, R., Ghoca, M., & Kleist, S. (1968). Alarm pheromone in the earthworm *Lumbricus terrestris. Science, 161*, 597–599.

Reynierse, J. H., & Ratner, S. C. (1964). Acquisition and extinction in the earthworm, *Lumbricus terrestris. Psychological Record, 14*, 383–387.

Reynierse, J. H., & Walsh, G. L. (1967). Behavior modification in the protozoan, *Stentor*, re-examined. *Psychological Record, 17*, 161–165.

Ricci, N. (1990). The behaviour of ciliated protozoa. *Animal Behaviour, 40*, 1048–1069.

Richards, R. J. (1987). *Darwin and the emergence of evolutionary theories of mind and behavior.* Chicago: The University of Chicago Press.

Richelle, M., & Lejeune, H. (1980). *Time in animal behaviour.* Oxford: Pergamon Press.

Ricker, J. P., Brzorad, J. N., & Hirsch, J. (1986). A demonstration of discriminative conditioning in the blow fly, *Phormia regina. Bulletin of the Psychonomic Society, 24*, 240–243.

Ricker, J. P., Hirsch, J., Holliday, M. J., & Vargo, M. A. (1986). An examination of claims for classical conditioning as a phenotype in the genetic analysis of *Diptera*. In J. L. Fuller & E. C. Simmel (Eds.), *Perspectives in behavioral genetics* (pp. 155–200). Hillsdale, NJ: Lawrence Erlbaum Associates.

Riesen, A. H. (1960). Learning. In R. H. Waters, D. A. Rethlingshafer, & W. E. Caldwell (Eds.), *Principles of Comparative Psychology* (pp. 177–207). New York: McGraw-Hill.

Roberts, M. B. V. (1962). The giant fiber reflex of the earthworm, *Lumbricus terrestris L.* I. The rapid response. *Journal of Experimental Biology, 39*, 219–227.

Roger, M., & Galeano, C. (1977). Activity of the crayfish caudal photoreceptor submitted to a conditioning paradigm. *Brain Research, 124*, 449–456.

Rosenkoetter, J. S., & Boice, R. (1975). Earthworm pheromones and T-maze performance. *Journal of Comparative and Physiological Psychology, 88*, 904–910.

Rosenzweig, M. R., & Leiman, A. L. (1989). *Physiological psychology* (2nd ed.). New York: Random House.

Rubadeau, D. O., & Conrad, K. A. (1963). An apparatus to demonstrate and measure operant behavior of arthropoda. *Journal of the Experimental Analysis of Behavior, 6*, 429–430.

Rushforth, N. B. (1965). Behavioural studies of the coelenterate *Hydra pirardi* Brien. *Animal Behaviour, 13* (Supplement 1), 30–42.

Rushforth, N. B. (1967). Chemical and physical factors affecting behavior in *Hydra*: Interactions among factors affecting behavior in *Hydra*. In W. C. Corning & S. C. Ratner (Eds.), *Chemistry of learning* (pp. 369–390). New York: Plenum.

Rushforth, N. B. (1973). Behavioral modifications in coelenterates. In W. C. Corning, J. A. Dyal, & A. O. D. Willows (Eds.), *Invertebrate learning: Vol. 1. Protozoans through annelids* (pp. 123–169). New York: Plenum.

Sahley, C. L. (1984). Behavior theory and invertebrate learning. In P. Bateson & P. Marler (Eds.), *Biology of learning* (pp. 181–196). Berlin: Springer-Verlag.

Sahley, C. L., Gelperin, A., & Rudy, J. W. (1981). One-trial learning modifies food odor preferences of a terrestrial mollusc. *Proceedings of the National Academy of Science, 78*, 640–642.

Sahley, C. L., & Ready, D. F. (1985). Associative learning modifies two behaviors in the leech, *Hirudo medicinalis. Society of Neuroscience Abstracts, 11*, 367.

Sahley, C. L., & Ready, D. F. (1988). Associative learning modifies two behaviors in the leech, *Hirudo medicinalis. Journal of Neuroscience, 8*, 4612–4620.

Sahley, C., Rudy, J. W., & Gelperin, A. (1981). An analysis of associative learning in a terrestrial mollusc. I. Higher-order conditioning, blocking, and a transient US preexposure effect. *Journal of Comparative Physiology, 144*, 1–8.

Sanders, C. J. (1985). Flight speed of male spruce budworm moths in a wind tunnel at different wind speeds and at different distances from a pheromone source. *Physiological Entomology, 10*, 83–88.

Sanders, C. J., Lucuik, G. S., & Fletcher, R. M. (1981). Responses of male spruce budworm to different concentrations of sex pheromone as measured in a sustained flight wind tunnel. *Canadian Entomologist, 113*, 943–948.

Sammataro, D., & Avitabile, A. (1978). *The beekeeper's handbook*. Dexter, MI: Peach Mountain Press, Ltd.

Sarnat, H. B., & Netsky, M. G. (1985). The brain of the planarian as the ancestor of the human brain. *Canadian Journal of Neurological Sciences, 12*, 296–302.

Schick, K. (1971). Operants. *Journal of the Experimental Analysis of Behavior, 15*, 413–423.

Schleidt, W. M., Yakalis, G., Donnelly, M., & McGarry, J. (1984). A proposal for a standard ethogram, exemplified by an ethogram of the blue breasted quail (*Coturnix chinensis*). *Zeitschrift für Tierpsychologie, 64*, 193–220.

Schneirla, T. C. (1929). Learning and orientation in ants. *Comparative Psychology Monographs, 6*, 1–143.

Schneirla, T. C. (1933). Motivation and efficiency in ant learning. *Journal of Comparative Psychology, 15*, 243–266.

Schneirla, T. C. (1949). Levels in the psychological capacities of animals. In R. W. Sellars, V. J. McGill, & M. Farber (Eds.), *Philosophy for the future: The quest of modern materialism* (pp. 243–286). New York: Macmillan.

Schneirla, T. C. (1950). The relationship between observation and experimentation in the field study of behavior. *Annual of the New York Academy of Sciences, 51*, 1022–1044.

Schneirla, T. C. (1953). Modifiability in insect behavior. In K. Roeder (Ed.), *Insect physiology* (pp. 723–747). New York: Wiley.

Schöne, H. (1961). Learning in the spiny lobster *Panulirus argus*. *Biological Bulletin, 121*, 354–365.

Schöne, H. (1984). *Spatial orientation: The spatial control of behavior in animals and man*. Princeton, NJ: Princeton University Press.

Schreurs, B. G. (1989). Classical conditioning of model systems: A behavioral review. *Psychobiology, 17*, 145–155.

Shepherd, G. M. (1988). *Neurobiology* (2nd ed.). Oxford: Oxford University Press.

Shishimi, A. (1985). Latent inhibition experiment with goldfish (*Carassius auratus*). *Journal of Comparative Psychology, 99*, 316–327.

Shorrocks, B. (1972). Drosophila. London: Ginn & Company.

Sigurdson, J. E. (1981a). Automated discrete-trials techniques of appetitive conditioning in honey bees. *Behavior Research Methods & Instrumentation, 13*, 1–10.

Sigurdson, J. E. (1981b). Measurement of consummatory behavior in honey bees. *Behavior Research Methods & Instrumentation, 13*, 308–310.

Simmel, E. C., & Ramos, F. (1965). Spatial probability learning in ants. *American Zoologist, 5*, 228 (abstract).

Simpson, S. J., & White, P. R. (1990). Associative learning and locust feeding: Evidence for a "learned hunger" for protein. *Animal Behaviour, 40*, 506–513.

Skinner, B. F. (1935). Two types of conditioned reflex and a pseudo type. *Journal of General Psychology, 12*, 66–77.

Skinner, B. F., & Barnes, T. C. (1930). The progressive increase in the geotropic response of the ant *Aphaenogaster*. *Journal of General Psychology, 4*, 102–112.

Smith, B. H. (1991). The olfactory memory of the honeybee *Apis mellifera*. I. Odorant modulation of short- and intermediate-term memory after single-trial conditioning. *Journal of Experimental Biology, 161*, 367–382.

Smith, B. H., & Abramson, C. I. (1992). Insect learning: Case studies in comparative psychology. In L. I. Nadel (Ed.), *Encyclopedia of learning and memory* (pp. 276–283). New York: Macmillan.

Smith, B. H., Abramson, C. I., & Tobin, T. R. (1991). Conditional withholding of proboscis extension in honeybees (*Apis mellifera*) during discriminative punishment. *Journal of Comparative Psychology, 105*, 345–356.

Smith, J. C., & Baker, H. D. (1960). Conditioning in the horseshoe crab. *Journal of Comparative and Physiological Psychology, 53*, 279–281.

Sokolov, V. A. (1959). Conditioned reflexes in snails *Physa acuta*. *Vestnik Leningrad University, 9*, 82–86.

Sokolove, P. G., Beiswanger, C. M., Prior, D. J., & Gelperin, A. (1977). A circadian rhythm in the locomotor behavior of the giant garden slug *Limax maximus*. *Journal of Experimental Biology, 66*, 47–64.

Solomon, R. L. (1977). An opponent-process theory of motivation. V. Affective dynamics of eating. In L. M. Barker, M. R. Best, & M. Domjan (Eds.), *Learning mechanisms in food selection* (255–269). Waco, TX: Baylor University Press.

Spence, K. W. (1956). *Behavior theory and conditioning*, New Haven, CT: Yale University Press.

Squire, L. R. (Ed.). (1992). *Encyclopedia of learning and memory*. New York: Macmillan.

Staddon, J. E. R., & Bueno, J. L. O. (1991). On models, behaviorism and the neural basis of learning. *Psychological Science, 2*, 3–11.

Staddon, J. E. R., & Zhang, Y. (1989). Response selection in operant learning. *Behavioral Processes, 20*, 189–197.

Stafstrom, C. E., & Gerstein, G. L. (1977). A paradigm for position learning in the crayfish claw. *Brain Research, 134*, 185–190.

Stoller, D., & Sahley, C. L. (1985). Habituation and sensitization of the shortening reflex in the leech *Hirudo medicinalis*. *Society of Neuroscience Abstracts, 11*, 367.

Stratton, L. O., & Coleman, W. P. (1973). Maze learning and orientation in the fire ant *Solenopsis saevissima*. *Journal of Comparative and Physiological Psychology, 83*, 7–12.

Stretton, A. O. W., & Kravitz, E. A. (1973). Intracellular dye injection: The selection of a procion yellow and its application in preliminary studies of neuronal geometry in the lobster nervous system. In S. B. Kater & C. Nicholson (Eds.), *Intracellular staining in neurobiology* (pp. 21–40). New York: Springer-Verlag.

Swartz, R. D. (1929). Modification of behavior in earthworms. *Journal of Comparative Psychology, 9*, 17–33.

Szentesi, A., & Bernays, E. A. (1984). A study of behavioral habituation to a feeding deterrent in nymphs of *Schistocerca gregaria*. *Physiological Entomology, 9*, 329–340.

Szlep, R. (1964). Change in the response of spiders to reported web vibration. *Behaviour, 23*, 203–239.

Szymanski, J. S. (1912). Modification of the innate behavior of cockroaches. *Journal of Animal Behavior, 2*, 81–90.

Tartar, V. (1961). *The biology of* Stentor. New York: Pergamon.

Tavolga, W. N. (1969). *Principles of animal behavior*. New York: Harper & Row.

Taylor, R. C. (1971). Instrumental conditioning and avoidance behavior in the crayfish. *Journal of Biological Psychology, 13*, 36–41.

Terrace, H. S. (1973). Classical conditioning. In J. A. Nevin (Ed.), *The study of behavior:*

Learning, motivation, emotion, and instinct (pp. 71–112). Glenview, IL: Scott, Foresman and Co.

Tempel, B. L., Bonini, N., Dawson, D. R., & Quinn, W. G. (1983). Reward learning in normal and mutant *Drosophila*. *Proceedings of the National Academy of Science*, *80*, 1482–1486.

Tesauro, G. (1986). Simple neural models of classical conditioning. *Biological Cybernetics*, *55*, 187–200.

Teyler, T. J. (1984). *A primer of psychobiology: Brain and behavior* (2nd ed.). New York: W. H. Freeman.

Teyler, T. J., Baum, W. M., & Patterson, M. M. (Eds.). (1975). Behavioral and biological issues in the learning paradigm. *Physiological Psychology*, *3*, 65–72.

Thomas, R. K. (1980). Evolution of intelligence: An approach to its assessment. *Brain, Behavior and Evolution*, *17*, 454–472.

Thompson, R., & McConnell, J. V. (1955). Classical conditioning in the planarian, *Dugesia dorotocephala*. *Journal of Comparative and Physiological Psychology*, *48*, 65–68.

Thompson, R. F., & Spencer, W. A. (1966). Habituation: A model phenomenon for the study of neuronal substrates of behavior. *Psychological Review*, *73*, 16–43.

Thon, B. (1987). Acquisition and retention of habituation as a function of intertrial interval duration during training in the blowfly. *Behavioural Processes*, *15*, 47–57.

Thon, B., & Pauzie, A. (1984). Differential sensitization, retention, and generalization of habituation in the blowfly (*Calliphora vomitoria*). *Journal of Comparative Psychology*, *98*, 119–130.

Thorndike, E. L. (1911). *Animal intelligence*. New York: Macmillan.

Thorpe, W. H. (1979). *The origins and rise of ethology*. London: Heinemann Educational.

Tinbergen, N. (1951). *The study of instinct*. Oxford: Clarendon Press.

Tinbergen, N. (1963). On aims and methods of ethology. *Zeitschrift für Tierpsychologie*, *20*, 410–433.

Tobach, E. (1987). *Historical perspectives and the international status of comparative psychology*. Hillsdale, NJ: Lawrence Erlbaum Associates.

Tolman, E. C. (1949). There is more than one kind of learning. *Psychological Review*, *55*, 189–208.

Tosney, T., & Hoyle, G. (1977). Computer-controlled learning in a simple system. *Proceedings of the Royal Society of London. Series B.*, *195*, 365–393.

Towne, W. F., & Kirchner, W. H. (1989). Hearing in honey bees: Detection of air-particle oscillations. *Science*, *244*, 286–288.

Townsend, J. C. (1953). *Introduction to experimental method*. New York: McGraw-Hill.

Tryon, R. C. (1930). Studies in individual differences in maze ability. I. The measurement of the reliability of individual differences. *Journal of Comparative Psychology*, *11*, 145–170.

Tully, T. (1984). *Drosophila* learning: Behavior and biochemistry. *Behavior Genetics*, *14*, 527–557.

Tully, T. (1991). Genetic dissection of learning and memory in *Drosophila melanogaster*. In J. Madden IV. (Ed.), *Neurobiology of learning, emotion and affect* (pp. 29–66). New York: Raven Press.

Tully, T., & Quinn, W. G. (1985). Classical conditioning and retention in normal and mutant *Drosophila melanogaster*. *Journal of Comparative Physiology A*, *157*, 263–277.

Tulving, E. (1985). On the classification problem in learning and memory. In L. Nilsson & T. Archer (Eds.), *Perspectives on learning and memory* (pp. 67–94). Hillsdale, NJ: Lawrence Erlbaum Associates.

Turner, C. H. (1892a). Psychological notes upon the gallery spider—illustrations of intelligent variations in the construction of the web. *Journal of Comparative Neurology*, *2*, 95–110.

Turner, C. H. (1892b). A few characteristics of the avian brain. *Science*, *19*, 16–17.

Turner, C. H. (1912). An experimental investigation of an apparent reversal of the response to light in the roach (*Periplaneta orientalis*, L). *Biological Bulletin*, *23*, 371–386.

Turner, C. H. (1914). An experimental study of the auditory powers of the giant silkworm moth (*Saturnidae*). *Biological Bulletin*, *27*, 325–332.

van der Steen, W. J., & Maat, A. (1979). Theoretical studies on animal orientation. I. Methodological appraisal of classifications. *Journal of Theoretical Biology*, *79*, 223–234.

VanDeventer, J. M., & Ratner, S. C. (1964). Variables affecting frequencies of response of planarian to light. *Journal of Comparative and Physiological Psychology*, *57*, 407–411.

Vargo, M., Holliday, M., & Hirsch, J. (1983). Automatic stimulus presentation for the proboscis extension reflex in Diptera. *Behavior Research Methods and Instrumentation*, *15*, 1–4.

Vogt, P. (1969). Dressur von sammelbienen auf sinusforming moduliertes flimmerlicht [Training of foragers to sinewave modulated flickering light]. *Zeitschrift für Vergleichende Physiologie*, *63*, 182–203.

von Frisch, K. (1914). Der farbensinn und formensinn der bienen [Color and shape vision in bees]. *Zoologische Jahrbücher*, *35*, 1–182.

von Frisch, K. (1962). Dialects in the language of bees. *Scientific American*, *207*, 79–87.

von Frisch, K. (1967). *The dance language and orientation of bees*. Cambridge, MA: Belknap Press.

Vowles, D. M. (1964). Olfactory learning and brain lesions in the ant (*Formica rufa*). *Journal of Comparative and Physiological Psychology*, *58*, 105–111.

Vowles, D. M. (1965). Maze learning and visual discrimination in the wood ant (*F. rufa*). *British Journal of Psychology*, *56*, 15–31.

Vowles, D. M. (1967). Interocular transfer, brain lesions, and maze learning in the wood ant, *Formica rufa*. In W. C. Corning & S. C. Ratner (Eds.), *Chemistry of learning* (pp. 425–451). New York: Plenum.

Wagner, V. A. (1894). *L'industrie des Araneina: Description systématique des constructions des Araignées* [The industry of spiders: A systematic description of the constructions of spiders]. St. Petersburg: Académie Impériale de Science.

Walker, I. (1972). Habituation to disturbance in the fiddler crab (*Uca annulipes*) in its natural environment. *Animal Behaviour*, *20*, 139–146.

Walker, M. M., Baird, D. L., & Bitterman, M. E. (1989). Failure of stationary but not flying honeybees (*Apis mellifera*) to respond to magnetic field. *Journal of Comparative Psychology*, *103*, 62–69.

Walters, E. T., Carew, T. J., & Kandel, E. R. (1979). Classical conditioning in *Aplysia californica*. *Proceedings of the National Academy of Sciences*, *76*, 6675–6679.

Warden, C. J., Jenkins, T. N., & Warner, L. H. (1935). *Comparative psychology—A comprehensive treatise: Vol. 1. Principles and methods*. New York: Ronald Press.

Warden, C. J., Jenkins, T. N., & Warner, L. H. (1940). *Comparative psychology—A comprehensive treatise: Vol. 2. Plants and invertebrates*. New York: Ronald Press.

Wasserman, E. A. (1989). Pavlovian conditioning: Is temporal contiguity irrelevant? *American Psychologist*, *44*, 1550–1551.

Wasserman, G. S., & Patton, D. G. (1969). Avoidance conditioning in *Limulus*. *Psychnomic Science*, *15*, 143.

Waterman, T. H. (Ed.). (1960). *The physiology of crustacea: Vol. 1. Metabolism and growth.* New York: Academic Press.

Waterman, T. H. (Ed.). (1961). *The physiology of crustacea: Vol. 2. Sense organs, integration and behavior.* New York: Academic Press.

Waters, R. H. (1960). The nature of comparative psychology. In R. H. Waters, D. A. Rethlingshafer, & W. E. Caldwell (Eds.), *Principles of comparative psychology* (pp. 1–17). New York: McGraw-Hill.

Watson, J. B. (1914). *Behavior: An introduction to comparative psychology.* New York: Holt.

Wayner, M. J., & Zellner, D. K. (1958). Role of the suprapharyngeal ganglion in spontaneous alternation and negative movements in *Lumbricus terrestris* L. *Journal of Comparative and Physiological Psychology, 51,* 282–287.

Wells, M. J. (1967). Sensitization and evolution of associative learning. In J. Salanki (Ed.), *Neurobiology of invertebrates* (pp. 391–411). Budapest: Academic Kiado.

Wells, P. H. (1973). Honey bees. In W. C. Corning, J. A. Dyal, & A. O. D. Willows (Eds.), *Invertebrate learning: Vol. 2. Arthropods and gastropod mollusks* (pp. 173–185). New York: Plenum.

Welsh, J. H. (1934). The caudal photoreceptor and responses of the crayfish to light. *Journal of Cell Comparative Physiology, 4,* 379–388.

West, L. S. (1951). *The housefly.* Ithaca, New York: Comstock Publishing Company.

Westerman, R. A. (1963a). Somatic inheritance of habituation of responses to light in planarians. *Science, 140,* 676–677.

Westerman, R. A. (1963b). A study of habituation of responses to light in the planarian *Dugesia dorotocephala. Worm Runners' Digest, 5,* 6–11.

Wharton, D. A. (1986). *A functional biology of nematodes.* Baltimore: John Hopkins University Press.

Wichterman, R. (Ed.). (1986). *The biology of paramecium* (2nd ed.). New York: Plenum.

Wickens, D. D., & Wickens, C. D. (1942). Some factors related to pseudoconditioning. *Journal of Experimental Psychology, 31,* 518–526.

Wight, K., Francis, L., & Eldridge, D. (1990). Food aversion learning in the hermit crab *Pagurus granosimanus. Biological Bulletin, 178,* 205–209.

Wilkens, L. A. (1988). The crayfish caudal photoreceptor: Advances and questions after the first half century. *Comparative Biochemistry and Physiology, 91C,* 61–68.

Willner, P. (1978). What does the headless cockroach remember? *Animal Learning and Behavior, 6,* 249–257.

Willows, A. O. D. (1971). Giant brain cells in mollusks. *Scientific American, 224,* 69–75.

Wilson, D. M., & Hoy, R. R. (1968). Optomotor reaction, locomotory bias, and reactive inhibition in the milkweed bug *Oncopeltus* and the beetle *Zophobas. Zeitschrift für vergleichende Physiologie, 58,* 136–152.

Wolcott, T. G., & Hines, A. H. (1989). Ultrasonic biotelemetry of muscle activity for free-ranging marine animals: A new method for studying foraging by blue crabs (*Callinectes sapidus*). *Biological Bulletin, 176,* 50–56.

Wood, D. C. (1970a). Electrophysiological studies of the protozoan, *Stentor coeruleus. Journal of Neurobiology, 1,* 363–377.

Wood, D. C. (1970b). Parametric studies of the response decrement produced by mechanical stimuli in the protozoan, *Stentor coeruleus. Journal of Neurobiology, 1,* 345–360.

Wood, D. C. (1971). Electrophysiological correlates of the response decrement produced by mechanical stimuli in the protozoan, *Stentor coeruleus. Journal of Neurobiology, 2,* 1–11.

Wood, D. C. (1972). Generalization of habituation between different receptor surfaces of *Stentor*. *Physiology & Behavior*, *9*, 161–165.

Wood, D. C. (1973). Stimulus specific habituation in a protozoan. *Physiology & Behavior*, *11*, 349–354.

Woods, P. J. (1974). A taxonomy of instrumental conditioning. *American Psychologist*, *29*, 584–596.

Woodworth, R. S. (1958). *Dynamics of behavior*. New York: Holt, Rinehart & Winston.

Wyers, E. J., Peeke, H. V. S., & Herz, M. J. (1964). Partial reinforcement and resistance to extinction in the earthworm. *Journal of Comparative and Physiological Psychology*, *57*, 113–116.

Wyers, E. J., Peeke, H. V. S., & Herz, M. J. (1973). Behavioral habituation in invertebrates. In H. V. S. Peeke & M. J. Herz (Eds.), *Habituation: Vol. 1. Behavioral studies* (pp. 1–57). New York: Academic Press.

Wyers, E. J., Smith, G. E., & Dinkes, I. (1974). Passive avoidance learning in the earthworm (*Lumbricus terrestris*). *Journal of Comparative and Physiological Psychology*, *86*, 157–163.

Yerkes, R. (1902). Habit formation in the green crab. *Biological Bulletin*, *3*, 241–244.

Young, J. Z. (1936). Structure of nerve fibers and synapses in some invertebrates. *Cold Spring Harbor Symposia on Quantitative Biology*, *4*, 1–5.

Zolman, J. F., & Peretz, B. (1987). Motor neuronal function in old *Aplysia* is improved by long-term stimulation of the siphon/gill reflex. *Behavioral Neuroscience*, *101*(4), 524–533.

Index

A

Abby-Kalio, N. J., 152

Abdominal ganglion, 184

Abramson, C. I., 165, 167

Activity wheel, 57–60, 227
  for earthworm, 57, 58
  for housefly, 57, 59, 60

Actograph, 60–62, 227. *See also* Activity wheel
  for crayfish, 62

Alkon, D., 15, 187

Alpha conditioning, 106, 128, 195, 204, 227–228

Amplitude of response, 228

Amsel, A., 194–195

Analogy, 228

Animal models, 24–25

Animals without Backbones (Buchsbaum et al.), 11–12

Annelids, 177–178

Anticipation method, 137

Ants
  complex maze for, 64, 65
  reward training for, 160–161
  T-maze performance of, 63

Aplysia
  classical conditioning of, 100, 186–187, 195–196
  habituation and sensitization in, 106, 184–186
  lever-press box for, 73
  neural research on, 184–187
  trace conditioning in, 139

Apparatus, research, 35, 40–41. *See also* Learning apparatus

Appetitive (classical reward) conditioning, 39, 135

Arthropods, 178

Associative learning, 38, 228
  cellular models of, 186–188

Automated problem box. *See* Lever-press box

Autoshaping, 125, 128

Aversive (defensive) conditioning, 39, 135

Avoidance conditioning, 199–202, 205, 228. *See also* Signaled avoidance conditioning

Avoidance training, 164

B

Backward conditioning, 139, 228

Base rate of responding
  classical conditioning, 130
  habituation/sensitization, 108
  instrumental and operant conditioning, 157–158

Bayley H., 203

Bees, 165
  free-flying procedure with, 86–87, 93
  inhibitory classical conditioning of, 122, 144
  lever-press box for, 70
  punishment training with, 161
  shuttle box for, 76, 80
  signaled avoidance conditioning of, 56

About the Author

Charles I. Abramson is an assistant professor in the Department of Psychology at Oklahoma State University (OSU) and was named Oklahoma State University College of Arts and Sciences 1994 Teacher of the Year and Oklahoma State University Department of Psychology Outstanding Teacher 1993–1994. Abramson has published empirical studies, review articles, and book chapters concerning various aspects of invertebrate learning. He is the author of *Invertebrate Learning: A Laboratory Manual and Source Book*, (American Psychological Association, 1990), and he edited (with Z. Shuranova and Y. Burmistrov) the forthcoming *Russian Contributions to Invertebrate Learning* (Praeger). Over his career, he has studied many types of invertebrates, including ants, bees, crabs, earthworms, flies, planarians, and protozoans. His research has appeared in the *Journal of Comparative Psychology*, *Physiology & Behavior*, *Journal of Neuroscience*, *Behavioral and Neural Biology*, *Animal Learning & Behavior*, *Biological Bulletin*, and the *Journal of the Experimental Analysis of Behavior*. He serves on the editorial boards of the *Journal of Mind and Behavior* and *Man, Neuron, Model: E-mail Communications in Psychophysiology* (a journal devoted to making Russian scientific work available to the West) and as consulting reviewer for a number of scientific journals.

Abramson received his AB in psychology from Boston University in 1978 and his PhD in experimental–physiological psychology from the same institution in 1986. After graduation, he spent several years as research assistant professor of biochemistry at the State University of New York Health Science Center in Brooklyn before moving to OSU. Abramson's research and teaching interests span several areas, including the comparative analysis of behavior, simple systems analysis of invertebrate learning, apparatus design, and the use of experimentation with invertebrates as a teaching tool in the classroom.